Newton's
Forgotten
Lunar
Theory

ii

Nicholas Kollerstrom

Newton's Forgotten Lunar Theory

His Contribution to
the Quest for
Longitude

Includes the complete text of Isaac Newton's
Theory of the Moon's Motion
1702

Green Lion Press

Santa Fe, New Mexico

Manufactured in the United States of America.

Published by Green Lion Press,
1611 Camino Cruz Blanca, Santa Fe, New Mexico 87501 USA.

Telephone (505) 983-3675; FAX (505) 989-9314;
Orders (within USA only) toll-free 1-888-636-8616.

mail@greenlion.com
www.greenlion.com

Green Lion Press books are printed on acid-free paper. Both softbound and cloth-bound editions have sewn bindings designed to lie flat and allow heavy use by students and researchers. Clothbound editions meet the guidelines for permanence and durability of the Committee on Production Guidelines for Book Longevity of the Council on Library Resources.

Printed and bound by Sheridan Books, Inc., Chelsea, Michigan.

Dust Jacket design and cover stamp based on William Crabtree's illustration of the Horroxian lunar theory, on which Newton's was based (see pp. 81-84). Design by William H. Donahue and Dana Densmore with help from Nadine Shea.

Cataloging-in-Publication Data:

Kollerstrom, Nicholas
Newton's forgotten lunar theory: his contribution to the quest for longitude / by Nicholas Kollerstrom.

Includes the complete text of Isaac Newton's *Theory of the Moon's Motion* of 1702, with introduction, commentary, glossary, extensive bibliography, and index.

ISBN 1-888009-08-X (cloth binding with dust jacket)

1. Newton, Isaac, works. 2. Lunar theory. 3. History of astronomy.
4. History of Science. 5. Longitude.
I. Nicholas Kollerstrom (1946-). II. Newton, Isaac (1642-1727). III. Title.

QB391.N4 K65 2000

Library of Congress Catalog Card Number 99-068025

Contents

Foreword by Curtis Wilson   xi

Author's Preface by Nicholas Kollerstrom   xv

Part I:

Text of Isaac Newton's Theory of the Moon's Motion   3

Part II: *Commentary and Analysis*

Chapter 1
  Introduction   17
    I. The Problem   17
    II. Accuracy of the "Theory"   20
    III. The Two Clock Method   22
    IV. Differing Approaches to the Problem   26
    V. The Apse Line Motion   27
    VI. Enlightenment Reception of TMM   29

Chapter 2
  Lunar Inequalities and Apse Motion   31
    I. Stages of Development   31
    II. The Horroxian Ellipse   32
    III. The Rocking Apse   33
    IV A Varying Eccentricity   36

Chapter 3
  Some Perspectives on TMM   41
    I. From Minutes to Seconds   41
    II. Conditions of Composition   42
    III. Halley's Hope   44
    IV. Moving the Goalposts   47
    V. A Modern Approach   48

Chapter 4
  A Commentary on TMM   51

Chapter 5
  Mean Motions of the Sun and Moon   57
  I.   The Modern Equations   58
  II. Newtonian Values   60
  III. A Century of Mean Motions   63
  IV. The Missing Flamsteed tables   65
  V.   Mean Motion Graphs   66
  VI. Summary   69

Chapter 6
  Finding the Anomaly   71
  I.   The solar anomaly   71
  II   Kepler Motion and the Equation of Centre   72
  III. The Lunar Equation of Centre   77
  IV   Finding the Prostaphaeresis   78

Chapter 7
  The Horrox-Wheel in Motion   81
  I.   A variable eccentricity   81
  II. TMM's Diagram   84
  III. Offbeat Octants   87
  IV. Halley's Contribution   89
  V.   Linkage of e and δ   93
  VI. Not the Evection   95

Chapter 8
  The Seven Moons of TMM   97
  I.   TMM in Machine-Readable Form   99
  II. The Seven Steps as trigonometric functions   105
  III. A Comparison with Flamsteed   106
  IV. A Test of Accuracy   109
  V.   Halley's Judgement   110

Chapter 9

Commentary on TMM, Continued 113
  I. An English Pamphlet 113
  II. The Annual Equations 113
  III. Two New Equations 115
  IV. The Horrox-Wheel Mechanism 118
  V. Amplitude of the Variation 121
  VI. A Second Horrox-Wheel 126
  VII. Latitude 128
  VIII. An Early Draft of TMM? 128

Chapter 10

Testing the Sevenfold Chain 133
  I. Five Historic Case-Studies 133
  II. The Erronous Sixth 137
  III. Newton's New Equations 137
  IV. Comparison with modern terms 139
  V. "Apogee in ye Summer Signs" 142
  VI. Testing the new equations 143
  VII. Comparison with DOS 145
  VIII. The First Computation 149
  IX. Some Conclusions 150

Chapter 11

Error Patterns 153
  I. The Error Envelope 153
  II. Error-Periods 155
  III. The "Hidden Terms" of Longitude 157
  IV. Latitude 159
  V. Flamsteed's View of DOS Errors 161

Chapter 12

TMM in the Principia 165
  I. Cotes' Contribution 166
  II. An Improvement on TMM? 168

III. An Equation of Eccentricity   170
IV. Constructing the New Epicycle   172
V. The Sixth Equation   174
VI. The Seventh Equation   176
VII. The Truncated 1726 Version   178
VIII. No Baricentre Correction   178
IX. Adding the Epicycle   179

Chapter 13

Halley and the Saros Synchrony   183
I. The President's Proposal   186
II. The Accuracy of Halley's Method   189
III. The Misunderstanding of Halley's Method   193
IV. Syzygy Accuracy   196
V. Halley's Version of TMM   197
VI. A Silent Crisis   201

Chapter 14

Construction of Tables   205
I. Flamsteed   205
II. Delisle   208
III. Halley   209
IV. LeMonnier   211
V. The Node Equation   214
VI. Wright Claims a Longitude Prize   215
VII. Leadbetter   216
IX. Other Tables   218
X. China   221

Chapter 15

Conclusion   225
I. French Comment   230
II. The Silent Decades   231
III. The Saros   232
IV. End of the Horrocksian Era   233

Appendices
   I. Some Astronomical Constants   237
   II. The Equations of Mean Motion   237
   III. Textbooks consulted, 1650–1750   239
   IV. Mean Motions from Textbooks 1650–1750   240
   V. Glossary of Terms Used in Text   241

Bibliography   245

Index   255

Foreword

In the study before us Dr. Nicholas Kollerstrom throws new light on important and hitherto unresolved questions concerning Newton's lunar theory. This theory, which first appeared in both Latin and English versions in 1702, and then in a revised form in the second edition of the *Principia* of 1713, has from the start been puzzling to its readers. How did Newton arrive at it—for he gives only numerical results and calculational directions without describing how he came by them? How accurate was it, compared with observations made around 1700? How much, if at all, did it improve on the theory it was designed to replace—the theory published by Flamsteed in his Doctrine of the Sphere of 1681, itself a refurbishing of the lunar theory of Jeremiah Horrocks? (Francis Baily in 1835 gave the distinct impression that Newton's theory was little if at all better than Flamsteed's.) To what extent were the new, non-Horrocksian terms that Newton inserted into the theory *right*—that is, contributory to goodness of empirical fit? And how influential was the theory? Which astronomers, in the ensuing decades, applied it in the making of predictions and the construction of predictive tables?

That the theory has not been properly assessed till now may seem surprising. Could it not be compared with a more modern and accurate theory, term by term? As Kollerstrom's detailed commentary on the theory (an excellent introduction to it) makes clear, the answer is, not without a rather difficult and delicate preparation. Newton's theory is constructed differently from the modern theories. In the latter, the argument of each corrective term is a linear combination of mean motions—of the Moon, Sun, lunar apse, lunar node, and solar perigee. The corrective terms are numerous, and the total correction to the mean motion is their sum. In Newton's theory, by contrast, the mean motions initially adopted for the Moon, its apse, and node, are progressively modified through "steps of equation"; the successive corrections after the first are no longer simply equivalent to any single additive term whose argument is a linear combination of mean motions, or to any finite sum of such terms. Thus Newton's theory and the modern theory are not strictly comparable. To be sure, one might undertake by approximation to reduce Newton's theory into the form of the modern theory; d'Alembert in fact did this in the 1750s, but the accuracy of his reduction has never been assessed. For a reliable analytic, term-by-term comparison of Newton's theory with a good modern theory, the reduction would have to be carried out anew, and with the utmost care.

Meanwhile, Dr. Kollerstrom has answered many of the unanswered questions by a quite different route. The new tool he brings to bear is a computerized modern lunar ephemeris, accurate for historical times. With this he compares predictions drawn from Newton's theory in the 18th century. He thus demonstrates conclusively that Newton's theory is more accurate than any of its predecessors. To test the four small terms that Newton adds to his basically Horrocksian -style theory, Kollerstrom determines the effect on predictive accuracy of omitting each such term in succession; he finds that all four terms contribute measurably to predictive accuracy—the theory is the better for including them. By a similar test he shows that the modification of the Flamsteedian eccentricity that Newton introduced at Halley's suggestion improves accuracy to an even greater extent. Some other features of Newton's theory—his subjection of the apse and node to annual equations—prove to be ineffectual modifications.

Newton's correction of the several mean motions involved in the lunar theory appears to have been carried out with particular care. His value for the twenty-year advance in the Moon's mean longitude is a little shy of the correct value, and the resulting drift in longitude has to be discounted in comparing other aspects of the theory with the modern ephemeris. The Newtonian values for the mean motions are distinctive, and Kollerstrom employs them in identifying authors dependent on Newton. Tables based on Newton's theory were first published in the 1730s. On the Continent, the tables included by Pierre-Charles LeMonnier in his *Institutions astronomiques* of 1746 were Newtonian—derivative, it appears, from tables prepared by Flamsteed shortly after Newton's theory first appeared; how they came to be in LeMonnier's hands is unknown. Perhaps the last Newtonian tables to be published were Halley's, in 1749 (although the Chinese used a modified Newtonian theory well into the present century). After the middle of the eighteenth century, however, the lion's share of interest went to tables derived analytically, in the mode of the Leibnizian calculus.

One question that neither Kollerstrom nor anyone else has answered is how Newton derived the small terms in his theory. In the second edition of the *Principia*, Newton asserts that these terms were derived from the theory of gravity, but does not explain how. This assertion is also implicitly made by the anonymous editor of the 1702 English edition of Newton's theory (see pp. 3–4 below). Should the assertion be believed? It is conceivable that Newton derived them by a kind of harmonic analysis from the observations supplied by Flamsteed. To date, we remain clueless as to how he arrived at these terms.

Among the topics that Kollerstrom's study illuminates is the method proposed by Edmond Halley for solving the longitude problem. Halley's idea was that tables derived from Newton's theory should be compared with observations over a complete Saros cycle of some eighteen years; the errors, he hypothesized, would repeat in successive Saros cycles, and could thus be used in conjunction with the tables to determine accurately the Moon's longitude for any time. In 1720, on becoming Astronomer Royal, Halley set out to make the requisite comparisons over a full Saros cycle; by the time of his death in 1742 he had completed the task. His tables and the tabulated errors, however, were not published till 1749, and their application to the finding of longitude at sea was never pursued. Indeed, from astronomers and historians Halley's proposal has received little in the way of attention or respect. Kollerstrom shows that the method works, and with astonishing success, giving the Moon's longitude to within about one arcminute, an accuracy sufficient to merit the £10,000 prize established by Parliament in 1714.

In sum, Kollerstrom has resolved important and hitherto unresolved puzzles concerning Newton's lunar theory. He corrects the errors of prior historians (including the present writer) with respect to it. He provides measures of its accuracy and influence more detailed and accurate than anything available previously. These are decisive advances on which future researchers can rely.

Curtis Wilson
St. John's College
Annapolis, Maryland

Author's Preface

AT THE DAWN OF THE NEW CENTURY, there appeared the first two textbooks of Newtonian astronomy: Gregory's *Astronomiae Elementa* of 1702, and Whiston's *Praelectiones Astronomicae* of 1707. They were meant to challenge the Cartesian philosophy then being taught in the schools of England, and both contained the full text of Isaac Newton's *Theory of the Moon's Motion* (hereinafter referred to as "*TMM*"). Though occupying a mere five pages of Gregory's book, it formed an essential part of that challenge, for it purported to show that the new Newtonian philosophy had a practical and not merely theoretical significance.

At least one of these books definitely claimed that *TMM* had achieved what was then regarded as well-nigh impossible: showing how to predict the Moon's position in the sky well enough to be of service for finding longitude. French astronomical treatises in the opening decades of the eighteenth century struck a rather sceptical note over this claim, while British ephemerides-composers tended to regard *TMM* as a kind of Holy Grail: something which would render possible the production of what was most desired, a reliable lunar ephemeris, if only it could be rightly interpreted.

Few were the sailors who made grey hairs, as the saying went, in those days. As their ships sailed back, laden with chocolate from Africa or silks from India, they were as we still say today, quite "at sea" once land disappeared. Huge prizes were offered for any means to find longitude. And yet, Britain's two most distinguished astronomers of the time—Captain Edmond Halley and the Reverend John Flamsteed—had more or less diametrically opposed opinions as to the real value of *TMM*. The former claimed that it profoundly improved lunar prediction, while the latter averred that it gave no real improvement upon existing tables.

Historians of science have been reluctant to comment upon the matter. Bernard Cohen was not exaggerating when in 1975 he stated:

> this work [*TMM*] has hardly ever been discussed (or even referred to) in the literature concerning Newton or the history of astronomy (Cohen 1975, p. 1)

As the literature there alluded to is of no small volume, such an omission would tend to suggest that this brief work was hardly significant. On the other hand, *TMM* was frequently reprinted through the first half of the eighteenth century, suggesting that it was exerting some kind of influence. To what extent this was practical, or mythical, is the subject of our inquiry.

Just about everything except the authorship of *TMM* remains unsettled. Was *TMM* ever in fact used? If so, would its prescriptions have defined the much-sought lunar position, to anything resembling the claim made by its publisher?

The onward-rolling tercentenary process has now passed by the anniversary of the commencement at Greenwich in 1691 of the most accurate series of positional astronomy readings ever made, of Flamsteed's marriage, Newton's nervous breakdown, and then the collaboration between these two on the great endeavour, not without strife, a linking together of theory and practice. This is an appropriate time to re-evaluate some of the traditional myths surrounding the subject. Sadly, the year 1998 saw the closure of Britain's Royal Greenwich Observatory, signifying an inglorious end to 323 years of distinguished service to the nation. It is therefore appropriate that we should now look back at the *raison d'etre* for its establishment, when a seemingly impossible task faced Newton and Flamsteed, which led, one-and-a-half-centuries later, to longitude's zero-line being placed through Greenwich. The story may ultimately have been one of success, but it did not seem that way to the persons in our story. Newton, when he sat on the Board of Longitude, did not recommend any particular method as workable, and Flamsteed came to advocate the Moons-of-Jupiter method in the early 1680s, soon after being established as the "King's Observator." Much of the acrimony between them stemmed from the fact that they were not within sight of an achievable goal.

I discovered, while undertaking my PhD upon *TMM*, that it held a pivotal position in the evolution of Newton's lunar theory. His *Philosophiae Naturalis Principia Mathematica* (here alluded to as *PNPM*, or the *Principia*) took in its first edition a theoretical approach which was, as Flamsteed came to realise, of little or no practical value to astronomers of the time. Fifteen years later, *TMM* appeared as an antithesis: containing no theory—as we would nowadays use the term—it developed a set of procedures for operating a strange and unique hybrid creation, of a Keplerian ellipse plus epicycle motions. The important question to ask about this device is, did it work? A synthesis then appeared in the Second Edition of the *Principia* of 1713, in which Newton endeavoured to show that gravity theory could account for these motions. *TMM* entered into the greatly enlarged Book III of the *Principia*, complicating its reference-framework by requiring it to be both tropical (i.e., zodiac degrees, as astronomers in practice required) and sidereal (the inertial stellar framework that the theory required). Did he succeed? Could his gravity theory account for such? That question is far from central to our inquiry. I may have succeeded in resolving what others before me had failed to resolve, by sidestepping that question!

We here study Newton's answer to what was the greatest practical problem of the age: the finding of longitude. I have called this the forgotten side of Newton's lunar theory. That it still remains so is shown by the recent Harvard *Quest for Longitude* symposium (1993), which contains no hint either that Newton composed a practical recipe for finding longitude, or that this became the basis for many of the publications on the subject in the first half of the eighteenth century. Historians perennially confuse this practical quest with the weighty, three-body problems with which the *Principia* grappled, and which, decades later, came to bear fruit in the work of Euler and Tobias Meyer—through their use of the Leibnizian calculus.

Dava Sobel has vividly described how the clockmaker John Harrisson won the longitude prize. Of the competition between the two methods, clock versus moon, she wrote:

> In longitude determination, a realm of endeavour where nothing had worked for centuries, suddenly two rival approaches of apparently equal merit ran neck and neck. Perfection of the two methods blazed parallel trails of development down the decades from the 1730s to the 1760s. (Sobel 1995, p. 89)

The account here given follows one of these methods, up to 1750; by which time the Newtonian recipe, developed from the brilliant mind of the young Jeremiah Horrox in the early 1630s, had finished its useful life. It was Britain's first lunar "theory". Effectively, it was the beginning of British astronomy. It flowered in the Second Edition of *PNPM* and thereby spread across Europe, and was then replaced. It was replaced by what we would nowadays call a theory.

David Gregory, in publishing *TMM* in 1702, used the word "theory," and we should therefore apprehend that, in the seventeenth century, that word was used by astronomers in relation to a kinematic diagram as would produce the desired motion, and it is in this sense that it will here be used. In the period we deal with, the explanation or "theory" involved constructing a *picture,* of how wheels moved within wheels. For readers wishing to familiarise themselves with the topic, the essays by Curtis Wilson in the *General History of Astronomy* describe such theories as were current in seventeenth-century Europe; Newton himself used this word for his 1713 endeavour on the same subject, which used his law of gravity to account for the elliptical and epicyclic motions.

Historians of astronomy in the next millenium will surely take for granted the notion here developed, that a computer replica of a historical theory or model is required to establish its efficacy. Other programs are then required to look at the accuracy of the observational data involved. By thus calibrating the theories used in earlier times we progress beyond the arm-waving conjectures of an earlier generation of science historians. This does however require confidence in the accuracy of the modern programs used. When did such accuracy become available to historians? It became available in the 1990s, in the decade in which the present study was accomplished.

To support this claim, we point to the recently improved mean-motion values of Jean Meeus (1991), based on the most modern data; the improved Δt tables (F. R. Stephenson, 1995) for obtaining Ephemeris Time from Universal Time, and a precise Equation of Time program (Hughes et. al., 1989) from the UK's Nautical Almanac Office, as retains its accuracy through past centuries. The establishing of a "handshake" between modern programs and historical data forms a major thread of this treatise. We thereby find that Flamsteed was the first to use Kepler's second law in practice, by which we mean that his tables had no error, no residual difference, as compared with the values generated by the modern equation-of-centre.

The quest for longitude revolved around the finding of a reliable method for ascertaining lunar celestial longitude to somewhat less than one arcminute of accuracy. It has hitherto not been at all easy for historians to ascertain who achieved this. It may be that Halley did so. Collaborating with Bernard Yallop at the RGO's Nautical Almanac Office, we found Halley's data was a deal more accurate than had been supposed. Thereby we came to disagree with such experts as the late Commander Derek Howse, who expressed a more traditional view.

Our story concerns what Newton did, but also how this became forgotten, and a different account put in its place. Newton helped this process! The story here may begin with the third edition of the *Principia* of 1726, as being the only edition translated into English and thus tending to eclipse whatever was said in earlier versions. No recipe or sequence of steps occurs there (Prop. 35 of Book III), no statement that thereby lunar longitude may be found, but only an analysis in terms of gravity theory of the separated equations. The explanations given of the annual equation and Variation were indeed found credible by posterity. No reader of this text would guess that a method for finding the lunar longitude and latitude had existed, as was the source of the equations described. The science historian William

Whewell writing in the 1830s seems to have been the last to apprehend this matter.

Before D. T. Whiteside's 1976 account, given at the tercentenary of the founding of the Royal Greenwich Observatory, historians assumed that Newton had used his calculus methods for tackling this problem. An imaginary view of his lunar theory had developed, which tended to merge with what the Continental astronomers achieved in the mid-eigtheenth century (and thereby deprived them of credit). The present account is very much based upon the researches of Whiteside, though he twice advised me that what I was endeavouring to accomplish, in reconstructing *TMM*, could not be done. His account of Newton's "despairing" re-introduction of three epicycles into what is otherwise Keplerian orbit-theory, forms the basis of my story.

There are some histories of our subject-matter worth reading, but only in French. One thinks here of D'Alembert's *Recherches sur different points importans du Systeme du Monde* (Volume 1, 1754); M. Bailly's *Histoire de l'Astronomie Moderne* (Volume II, 1779); and M. Delambre's *Histoire de l'Astronomie au 18th Siecle* (1827). English-language histories of astronomy covering this period are not to be recommended, in part owing to a degree of mythologising to which they succumbed concerning the figure of Isaac Newton, but mainly because *TMM* has disappeared from view in these histories. The exceptions here are William Whewell's 1837 treatment of the subject, and that by Curtis Wilson above-mentioned.

The recently-published *Flamsteed's Stars* (Willmoth, 1997) has its eleven contributors looking at different aspects of the life-work of Britain's first Astronomer-Royal. Not one of them mentions his *Doctrine of the Sphere* (1681), as being the sole work that he published in his lifetime. No hint is given there of Whiteside's thesis, that Newton took over his "Horroxian" lunar theory from Flamsteed, while adding certain components to it. Flamsteed brought down from the Midlands the first British astronomical theory, whose career was to span more than a century.

Some interest was aroused by the findings here reported about Edmond Halley's use of the "Saros." He did rather more than give a name to this cycle and use it for solar eclipse prediction. Before his time, it remains equivocal as to whether the ancients had applied the term Saros for the interval of 18 years and 11 days, or used it for eclipse prediction. French historians such as Delambre accepted that Halley had been the first to use it for such. The first systematic use of *TMM* was by Halley in the context of the Saros-interval, over the 18 years 1722–40 for which he worked upon the lunar problem. The recent biography of Halley by Alan Cook has described how he used Flamsteed's data to compare with his own, over the Saros interval.

What, after all, is the Saros? It is a high-precision coincidence, between the several lunar-monthly cycles, and there are (I believe) no rational grounds as to why it should exist. The astronomer Fred Hoyle alluded to it as a "fluke" (Hoyle 1977, p. 100). It only becomes perceptible as an eclipse-predictor once eclipse data is available from different parts of the world and not just from a single location. Halley used the Saros periodicity to allow for the errors in Newton's *TMM*, since these returned according to the rhythm of the Saros. He realised that he could thereby attain a remarkable degree of accuracy, and dedicated his tenure as Astronomer Royal to achieving this goal. French astronomers were sceptical over Halley's method, because they could see no rationale to it. While in general the historian should not claim to know better than the historical characters about the key issues, the computer tells us that Halley's method worked, at least as well as he supposed it did. The computer does put us in a position to know better than the historical characters over matters of acuracy. That is a novelty of recent years.

This study aims to rescue Halley's reputation as Astronomer Royal in several respects. It explains properly and for the first time, what he was attempting, and affirms that his method worked. It further found that his observations were not inaccurate or unreliable, but on the contrary, as the first systematic observations using the new GMT were more accurate (as lunar-centre meridian-transits) than anything previously attained. On the negative side of the balance-sheet, Halley's publicly-made claims for the accuracy of *TMM* (doubtless expressing his loyalty to Newton) are found to be exaggerated. Also, Halley tinkered with *TMM*, simplifying it by stripping off some ancillary equations which he deemed to be superfluous, and thereby made it *less* accurate. The historical record remains silent concerning any endeavours of Halley to communicate his intent to to his contemporaries. In his advanced years, he faithfully followed the 24.8-hour lunar day through long years, sacrificing his sleep. We must presume that some hope of claiming the longitude prize motivated his secrecy, which resulted however in his method passing largely into oblivion.

TMM reached China, where it became accepted around 1740, in the context of a stationary Earth. It required the most subtle Jesuitical logic to explain to the Peking astronomers how a Keplarian-ellipse lunar orbit, sprinkled with a few epicycles, could move around a Tycho-Brahean immobile earth. The absence of gravitational theory in *TMM* was an advantage, in enabling the Chinese adoption of *TMM*! The Emperor took a dim view of western innovations; however, *TMM* enabled the Chinese, for the first time, to predict their eclipses reliably. In the seventeenth century, Chinese astronomers were still failing to predict solar eclipses! Whereas in the West *TMM* was required for the finding of longitude, in China this was not a vital issue. Rather, the state religion required the knowledge of solar-lunar astronomy to be able to predict eclipses. *TMM* remained in use in the Chinese state calendar right up into the nineteenth century. It there endured the transition into a heliocentric universe. Thereby it remained alive and well in China, long after Europe had forgotten it.

By way of assistance, the following books are recommended: the fragment Volume 2A of the *General History of Astronomy* (CUP), which series remains incomplete; Bernard Cohen's *Newton's Theory of the Moon's Motion*, a collector's item published back in 1975; Dava Sobel's *Longitude* (1995), and Volume II of *The Flamsteed Correspondence*. David Howse's essays on how the longitude method was applied, as in his chapter in the *Quest for Longitude* (Andrewes 1996), are helpful. Some readers may wish to download the relevant computer programs (in Excel format), available from the following website (constructed by the author):

http://www.ucl.ac.uk:80/sts/kollrstm/newton.htm.

(If this address ceases to be valid, you may either 1) search the University College, London, website (www.ucl.ac.uk) for Kollerstrom, or 2) go to the Green Lion Press website (www.greenlion.com) for an updated address.)

The programs simulate the lunar theories of the historical characters that concern us.

Acknowledgements

In the last few years of its life, Britain's Royal Greenwich Observatory looked back, to the work of its founder. The director of its Nautical Almanac Office, Bernard Yallop, collaborated with this writer in checking out John Flamsteed's observational data: the early observations of Flamsteed and Halley, the Equation of Time used by Flamsteed, and other key parameters. Without this the present inquiry would have been at sea. Dr Craig Waff kindly made available his collection of all the observatories and astronomers in eighteenth century Europe who had used *TMM*, collected from his visits to European archives. He was prevented from following through on his researches by a career switch (into a history of NASA's Deep Space exploration, which is surely just as interesting). His finding have been summarised by Bernard Cohen, 1975. For French response to *TMM*, especially Nicholas Delisle, I am obliged to Simon Schaffer for use of his manuscripts. For much guidance and advice and in pointing out what would have been major blunders I am obliged to Curtis Wilson. For advice on equations, and the providing of programs I thank Jean Meeus. Computer expert Guy Atkinson helped me to make my program work, that plotted the "error envelope" of *TMM*. Dr Derek McNally especially helped in encouragement and in discussing each chapter. Jeremiah Horrocks's diagram of his "crank" mechanism (Fig. 7.1) is reproduced by permission of the Syndics of Cambridge University Library, and the photograph of Flamsteed's table (Fig. 6.1, from J. Moore's *New System of Mathematicks*, Vol. I p. 96) is reproduced by permission of the Royal Astronomical Society. Reproduction rights are retained by the respective copyright holders. Finally, I thank William Donahue, my editor at Green Lion Press, for his excellent, helpful, and scholarly guidance in clarifying the text.

Abbreviations used in Text

Corr. – *The Correspondence of Isaac Newton*, especially Volume IV, 1694–1709, ed. J. F. Scott, Cambridge 1967.

DOS – "Doctrine of the Sphere" by John Flamsteed, published (anonymously) as *De Sphaera* in 1681, in Jonas Moore's *New Systeme of Mathematics*.

GHA – *The General History of Astronomy*, Volume 2A, Ed. René Taton and Curtis Wilson (Cambridge University Press, 1989). The reference will normally be to the two chapters by Curtis Wilson.

PNPM – Isaac Newton, *Principia Naturalis Philosophica Mathematica* (London, 1687). References are normally to the Second Edition of 1713. Page numbers refer to the variant edition edited by A. Koyré, I. Bernard Cohen, and Anne Whitman (2 Vols., Harvard University Press, 1972). The Third Edition of 1727 is available in a new translation by I. Bernard Cohen and Anne Whitman (University of California Press, 1999).

Phil. Trans. – *Philosophical Transactions of the Royal Society*.

TMM – *Theory of the Moon's Motion* by Isaac Newton, published in Latin by David Gregory in his *Astronomiae Elementa* of 1702. An English translation appeared in 1702 which is the text here referred to as *TMM*, reproduced by Cohen in 1975 and in Part I below.

TMM-PC – Computer-simulated model of the *TMM* procedure.

Part I

Text of the First English Edition
of Sir Isaac Newton's

Theory of the Moon's Motion

Text of Theory of the Moon's Motion

The text below is the first English edition, published in 1702 as a separate booklet. The Latin text had appeared in David Gregory's *Astronomiae Elementa* (Gregory 1702) shortly before. It is identical in content to the Latin except for a divergence in the seventh equation. The editor and author of the prefatory matter is not named, but is usually identified as Edmond Halley. The text itself may be Newton's original English rather than a translation of Gregory's version. For further details, see Ch. 9, I.

To the Reader

The Irregularity of the Moon's Motion hath been all along the just Complaint of Astronomers; and indeed I[1] have always look'd upon it as a great Misfortune that a Planet so near us as the Moon is, and which might be so wonderfully useful to us by her Motion, as well as her Light and Attraction (by which our Tides are chiefly occasioned) should have her Orbit so unaccountably various, that it is in a manner vain to depend on any Calculation of an Eclipse, a Transit, or an Appulse of her, tho never so accurately made. Whereas could her Place be but truly calculated, the Longitudes of Places would be found every where at Land with great Facility, and might be nearly guess'd at Sea without the help of a Telescope, which cannot be used.[2]

This Irregularity of the Moon's Motion depends (as is now well known, since Mr. Newton hath demonstrated the Law of Universal Gravitation)[3] on the Attraction of the Sun, which perturbs the Motion of the Moon (and of all other Satellites or Secondary Planets) and makes her move sometimes after and sometimes slower in her Orbit; and makes consequently an

---

[1] Probably Halley; see Ch. 9, I, below.

[2] For the importance of finding longitudes, see Ch. 1 Sect. 2 below.

[3] Universal gravitation was established in *PNPM* III Prop. 1–8, pp. 565–584, in 1687, and Newton had hoped soon thereafter to produce a gravitationally derived lunar theory, a hope that was never realized. See below, Ch. 1, III, and the scholium to the First Edition of *PNPM* III Prop. 35, p. 658.

3

Alteration in the Figure of that Orbit, as well as of its
Inclination to the Plain of the Ecliptick.[4] But this being now to
be accounted for, and reduced to a Rule; by this Theory such
Allowances are made for it, as that the Place of the Planet shall
be truly Equated.[5]

This therefore being perfectly New, and what the Lovers of
Astronomy have a long while been put in hopes to receive from
the Great Hand that hath now finished it; I thought it would be
of good service to our Nation to give it an English Dress, and
publish it by it self: For as Dr. Gregory's Astronomy is a large
and scarce Book, it is neither every ones Money that can pur-
chase it, nor Acquaintance that can procure it; and besides I
hope we have a great many Persons in England that have Skill
and Patience enough to calculate a Planet's Place, who yet it
may be don't well enough understand the Latin Tongue to
make themselves Masters of this Theory in the Author's own
Words. At least I persuade my self, that a Theory so easy and
plain as this, which carries along with it such Pretence to
Exactness, will encourage many Persons to imploy themselves
in Astronomical Calculation, which before possibly they neglected, because they judged there was but little Exactness to be
attained in it. And this would be a very useful way of spending
their leisure Hours; and if they would oblige us with the

---

[4] In the many corollaries to *PNPM* I Prop. 66 (pp. 278–294), Newton considered the
combined effects of the solar and terrestrial gravity on the moon in a hypothetical
and largely qualitative way. In Book III, where the mathematical conclusions of
Book I are applied to the world as it is observed, he devoted a long series of propositions (Prop. 22 and Props. 25–35, pp. 610–612 and 618–664) to an attempt at a
quantitative treatment of the moon's motions using gravitational principles.
Although he was able to deduce a few motions, such as the variation (*TMM* equation 5; see pp. 165–7 below), in the end he had to admit defeat, writing, "But we
also do not consider our computation to be accurate enough." (*PNPM* p. 658.)

[5] The "equated place" is the moon's true position (parallax and refraction being
ignored), which is computed from its mean position by the application of various
"equations." (See below, Ch. 6, especially Section II.) The claim being made here
is that the *TMM* equations were deduced from the dynamic principles developed
in the *Principia*. This claim is not supported by the evidence (see below, Ch. 15,
pp 228–9).

Publication of good Ephemerides, Tables, etc. they would soon enflame others with a Desire of pursuing these kind of Studies.

The Famous Mr. Isaac Newton's Theory of the Moon

This Theory hath been long expected by the Lovers of Art, and is now publish'd in Dr. Gregory's Astronomy, in Mr. Newton's own Words.

By this Theory, what by all Astronomers was thought most difficult and almost impossible to be done, the Excellent Mr. Newton hath now effected, viz., to determine the Moon's Place even in her Quadratures, and all other Parts of her Orbit, besides the Syzygys, so accurately by Calculation, that the Difference between that and her true Place in the Heavens shall scarce be two Minutes,[6] and is usually so small, that it may well enough be reckon'd only as a Defect in the Observation. And this Mr. Newton experienced by comparing it with the very many Places of the Moon observed by Mr. Flamstead, and communicated to him.

[Here the actual text of TMM *begins. A detailed commentary on the first seven paragraphs appears in Chapter 4 below, while the commentary on the remainder is in Chapter 9.]*

The Royal Observatory at Greenwich is to the West of the Meridian of Paris 2° 19′. Of Uraniburgh 12° 51′ 30″. And of Gedanum 18° 48′.[7]

---

[6] This claim of accuracy is excessive. This original version of *TMM* reversed the sign of the sixth equation, with the result that its predictions could err by as much as eight arcminutes (Figures 11.2 a–b). But even with the sixth equation used correctly, it was sometimes five minutes off, the standard deviation being 1.9 arcminutes. Nevertheless, the theory was substantially more accurate than its predecessors. See Ch. 10, IX and Ch. 11, I below.

[7] See the commentary in Ch. 4, p. 51.

The mean Motions of the Sun and Moon, accounted from the Vernal Equinox at the Meridian of Greenwich, I make to be as followeth. The last Day of December 1680. at Noon (Old Stile) the mean Motion of the Sun was 9Sign. 20° 34′ 46″. Of the Sun's Apogæum was 3 Sign. 7° 23′ 30″.[8]

The mean Motion of the Moon at that time was 6 Sign. 1° 35′ 45″. And of her Apogee 8 Sign. 4° 28′ 5″. Of the Ascending Node of the Moon's Orbit 5 Sign. 24° 14′ 35″.[9]

And on the last Day of December 1700. At Noon, the mean Motion of the Sun was 9 Sign 20°. 43′.50″. Of the Sun's Apogee 3 Sign. 7°. 44′. 30″. The mean Motion of the Moon was 10 Sign. 15°. 19′. 50″. Of the Moon's Apogee 11 Sign. 8°.18′.20″. And of her ascending Node 4 Sign. 27°. 24′. 20″. For in 20 Julian Years of 7305 Days, the Sun's Motion is 20 Revol. 0 Sign. 0°. 9′. 4″. And the Motion of the Sun's Apogee 21′. 0″.[10]

The Motion of the Moon in the same Time is 247 Rev. 4 Sign. 13°. 34′. 5″. And the Motion of the Lunar Apogee is 2 Revol. 3 Sign. 3°. 50′. 15″. And the Motion of her Node 1 Revol 0 Sign. 26°. 50′. 15″.[11]

All which Motions are accounted from the Vernal Equinox: Wherefore if from them there be subtracted the Recession or Motion of the Equinoctial which is 16′. 0″. there will remain the Motions in reference to the Fixt Stars in 20 Julian Years; viz. The Sun's 19 Revol. 11 Sign. 29°. 52′. 24″. Of his Apogee 4′. 20″. and the Moon's 247 Revol. 4 Sign. 13°. 17′. 25″. Of her Apogee 2 Revol. 3 Sign. 3°. 33′. 35″. And of the Node of the Moon 1 Revol. 0 Sign. 27°. 6′. 55″.[12]

According to this Computation the *Tropical Year* is 365 Days. 5 Hours. 48′.57″. And the *Sydereal Year* is 365 Days. 6 Hours. 9′.14″.[13]

---

[8] See the commentary in Ch. 4, pp. 51–2

[9] A "node" is the apparent position of the intersection of the planes of the Moon's and Earth's orbits. There are two of them; the "ascending" node is the one at which the Moon is passing from south of the ecliptic to north of it. For further discussion, see the commentary in Ch. 4, pp. 52–3.

[10] See the commentary in Ch. 4, pp. 53–5.

[11] See the commentary in Ch. 4, p. 55.

[12] See the commentary in Ch. 4, pp. 55–6.

[13] See the commentary in Ch. 4, p. 56.

These mean Motions of the Luminaries are affected with various Inequalities: Of Which,

1. There are the Annual Equations of the aforesaid mean Motions of the Sun and Moon, and of the Apogee and Node of the Moon.

The Annual Equation of the mean Motion of the Sun depends on the Eccentricity of the Earth's Orbit round the Sun, which is $16\frac{11}{12}$ of Such Parts, as that the Earth's mean Distance from the Sun shall be 1000: Whence 'tis called the *Equation of the Centre*; and is when greatest 1°. 56′. 20″

The greatest Annual Equation of the Moon's mean Motion is 11′. 49″.of her Apogee 20′. And of her Node 9′. 30″.[14]

And these four Annual Equations are always mutually proportional one to another: Wherefore when any of them is at the greatest, the other three will also be greatest; and when any one lessens, the other three will also be diminished in the same *Ratio*.

The Annual Equation of the Sun's Centre being given, the three other corresponding Annual Equations will be also given; and therefore a Table of *That* will serve for all. For if the Annual Equation of the Sun's Centre be taken from thence, for any Time, and be called P, and let $\frac{1}{10} P = Q$, $Q + \frac{1}{60} Q = R$, $\frac{1}{6} P = D$, $D + \frac{1}{30} D = E$, and $D - \frac{1}{60} D = 2 F$; then shall the Annual Equation of the Moon's mean Motion for that time be R, that of the Apogee of the Moon will be E, and that of the Node F.

Only observe here, that if the Equation of the Sun's Centre be required to be *added*; then the Equation of the Moon's mean Motion must be *subtracted*, that of her Apogee must be *added*, and that of the Node *subducted*. And on the contrary, if the Equation of the Sun's Centre were to be *subducted*, the Moon's Equation must be *added*, the Equation of her Apogee *subducted*, and that of her Node *added*.[15]

There is also an *Equation of the Moon's mean Motion* depending on the Situation of her Apogee in respect of the Sun; which

---

[14] These four paragraphs are discussed in Ch. 9, II, pp. 113–4.
[15] These four paragraphs are discussed in Ch. 9, II, pp. 114–5.

is *greatest* when the Moon's Apogee is in an Octant with the Sun, and is nothing at all when it is in the Quadratures or Syzygys. This Equation, when greatest, and the Sun *in Perigaeo*, is 3'. 56". But if the Sun be *in Apogeo*, it will never be above 3'. 34". At other Distances of the Sun from the Earth, this Equation, when greatest, is reciprocally as the Cube of such Distance.

But when the Moon's Apogee is any where but in the *Octants*, this Equation grows less, and is mostly at the same distance between the Earth and Sun, as the Sine of the double Distance of the Moon's Apogee from the next Quadrature of Syzygy, to the Radius.

This is to be *added* to the Moon's Motion, while her Apogee passes from a Quadrature with the Sun to a Syzygy; but is to be *subtracted* from it, while the Apogee moves from the Syzygy to the Quadrature.[16]

There is moreover another *Equation of the Moon's Motion*, which depends on the Aspect of the Nodes of the Moon's Orbit with the Sun: and this is *greatest* when her Nodes are in *Octants* to the Sun, and vanishes quite, when they come to their Quadratures or Syzygys. This Equation is proportional to the Sine of the double Distance of the Node from the next Syzygy or Quadrature; and at greatest is but 47". This must be added to the Moon's mean Motion, while the Nodes are passing from their Syzygys with the Sun to their Quadratures with him; but subtracted while they pass from the Quadratures to the Syzygys.[17]

From the Sun's true Place take the equated mean Motion of the Lunar Apogee, as was above shewed, the Remainder will be the Annual Argument of the said apogee. From whence the *Eccentricity of the Moon*, and the *second Equation of her Apogee* may be compar'd after the manner following (*which takes place also in the Computation of any other intermediate Equations.*)[18]

Let *T* represent the Earth, *TS* a Right Line joining the Earth and Sun, *TACB* a Right Line drawn from the Earth to the

---

[16] The preceding three paragraphs are discussed in Ch. 9, III, pp. 115–7.

[17] This paragraph is discussed in Ch. 9, III, pp. 117–8.

[18] This paragraph is discussed in Ch. 9, IV, p. 118.

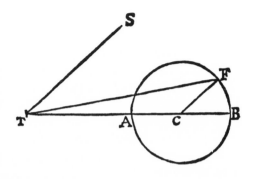

middle or mean Place of the
Moon's Apogee, equated,
as above: Let the Angle
STA be the Annual
Argument of the aforesaid
Apogee, *TA* the least
Eccentricity of the Moon's
Orbit, *TB* the greatest.
Bisect *AB* in *C*; and on the
Centre *C* with the Distance
AC describe a Circle *AFB*,
and make the Angle *BCF* =to the double of the Annual
Argument. Draw the Right Line *TF*, that shall be the
Eccentricity of the Moon's Orbit; and the Angle *BTF* is the
second Equation of the Moon's Apogee required.[19]

In order to whose Determination let the mean Distance of
the Earth from the Moon, or the Semidiameter of the Moon's
Orbit, be 1000000; then shall its *greatest Eccentricity TB* be 66782
such Parts; and the *least TA*, 43319. So that the greatest Equation
of the Orbit, viz. when the Apogee is in the Syzygys, will be 7°.
39′. 30″. or perhaps 7°. 40′. (for I suspect there will be some
Alteration according to the Position of the Apogee in Cancer or
in Capricorn.) But when it is in Quadrature to the Sun, the
greatest Equation aforesaid will be 4°. 57′. 56″. and the greatest
Equation of the Apogee 12°. 15′. 4″.[20]

Having from these Principles made a Table of the Equation
of the Moon's Apogee, and of the Eccentricitys of her Orbit to
each degree of the Annual Argument, from whence the
Eccentricity *TF*, and the Angle *BTF* (*viz*. The second and
principal Equation of the Apogee) may easily be had for any
Time required; let the Equation thus found be added, to the
first Equated Place of the Moon's Apogee, of the Annual
Argument be less than 90°, or greater than 180°, and less than
270°; otherwise it must be subtracted from it: and the Sum or
Difference shall be the Place of the Lunar Apogee secondarily

---

[19] This paragraph is discussed in Ch. 9, IV, p. 119.
[20] This paragraph is discussed in Ch. 9, IV, pp. 119–20.

equated; which being taken from the Moon's Place equated a third time, shall leave the mean Anomaly of the Moon corresponding to any given Time. Moreover, from this mean Anomaly of the Moon, and the before-found Eccentricity of her Orbit, may be found (by means of a Table of Equations of the Moon's Centre made to every degree of the mean Anomaly, and some Eccentricitys, *viz.* 45000, 50000, 55000, 60000, and 65000) the *Prostapharesis* or Equations of the Moon's Centre, as in the common way: and this being taken from the former Semicircle of the middle Anomaly, and added in the latter to the Moon's Place thus thrice equated, will produce the Place of the Moon a fourth time equated.[21]

The greatest Variation of the Moon (*viz.* That which happens when the Moon is in an Octant with the Sun) is, nearly, reciprocally as the Cube of the Distance of the Sun from the Earth. Let that be taken 37'. 25". when the Sun is *in Perigeo*, and 33'. 40". when he is *in Apogeo*: And let the Differences of this Variation in the Octants be made reciprocally as the Cubes of the Distances of the Sun from the Earth; and so let a Table be made of the aforesaid Variation of the Moon in her *Octants* (or its Logarithms) to every *Tenth*, *Sixth*, or Fifth Degree of the mean Anomaly: And for the Variation out of the Octants, make, as Radius to the Sine of the double Distance of the Moon from the next Syzygy or Quadrature :: so let the aforefound Variation in the Octant be to the Variation congruous to any other Aspect; and this *added* to the Moon's Place before found in the first and third Quadrant (accounting form the Sun) or *subtracted* from it in the second and fourth, will give the Moon's Place equated a fifth time.[22]

Again, as Radius to the Sine of the Sum of the Distances of the Moon from the Sun, and of her Apogee from the Sun's Apogee (or the Sine of Excess of that Sum above 360°.) :: so is 2'. 10". to a sixth Equation of the Moon's Place, which must be *subtracted*, if the aforesaid Sum or Excess be less than a

---

[21] For this paragraph, see Ch. 9, IV, pp. 120–1.

[22] For this paragraph, see Ch. 9, V, p. 122.

Semicircle, but *added*, if it be greater. Let it be made also, as Radius to the Sine of the Moon's Distance from the Sun :: so 2′. 20″.to a seventh Equation: which, when the Moon's light is encreasing, *add*, but when decreasing, *subtract*; and the Moon's place will be equated a seventh time, and this is her Place *in her proper Orbit.*[23]

Note here, the Equation thus produced by the mean Quantity 2′. 20″. is not always of the same Magnitude, but is encreated and diminished according to the Position of the Lunar Apogee. For if the Moon's Apogee be in Conjunction with the Sun's, the aforesaid Equation is about 54″. greater: but when the Apogees are in opposition,'tis about as much less; and it librates between its greatest Quantity 3′. 14″. and its least 1′. 26″. And this is when the Lunar Apogee is in Conjunction or Opposition with the Sun's: But in the Quadratures the afore-said Equation is to be lessen'd about 50″. or one minute, when the Apogees of the Sun and Moon are in Conjunction; but if they are in Opposition, for want of a sufficient number of Observations, I cannot determine whether it is to be lessen'd or increas'd. And even as to the Argument or Decrement of the Equation 2′. 20″.above mentioned, I dare determine nothing certain, for the same Reason, *viz.* the want of Observation accurately made.[24]

If the sixth and seventh Equations are augmented or dimin-ished in a reciprocal *Ratio* of the Distance of the Moon from the Earth, *i.e.*, in a direct *Ratio* of the Moon's Horizontal Parallax; they will become more accurate: And this may readily be done, if Tables are first made to each Minute of the said Parallax, and to every Sixth or Fifth Degree of the Argument of the sixth Equation for the *Sixth*, as of the Distance of the Moon from the Sun, for the Seventh Equation.[25]

From the Sun's Place take the mean Motion of the Moon's ascending Node, equated as above; the Remainder shall be the Annual Argument of the Node, whence its second Equation

---

[23] For this paragraph, see Ch. 9, V, pp. 123–4.

[24] See Ch. 9, V, pp. 124–5.

[25] See Ch. 9, V, p. 125.

may be computed after the following manner in the preceding Figure.

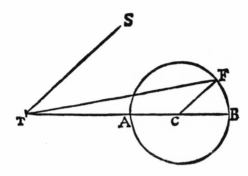

Let *T* as before represent the Earth, *TS* a right line conjoining the Earth and Sun: Let also the Line *TACB* be drawn to the Place of the Ascending Node of the Moon, as above equated; and let *STA* be the Annual Argument of the Node. Take *TA* from a Scale, and let it be to *AB* :: as 56 to 3, or as $18\frac{2}{3}$ to 1. Then bisect *BA* in *C*, and on *C* as a Centre, with the Distance *CA*, describe a Circle as *AFB*, and make the Angle *BCF* equal to double the Annual Argument of the Node before found: So shall the Angle *BTF* be the second Equation of the Ascending Node: which must be *added* when the Node is passing from a Quadrature to a Syzygy with the Sun, and *subducted* when the Node moves from a Syzygy towards a Quadrature. By which means the true Place of the Node of the Lunar Orbit will be gained: whence from Tables made after the common way, the *Moon's Latitude, and the reduction of her orbit to the Ecliptick*, may be computed, supposing the Inclination of the Moon's Orbit to the Ecliptick to be 4°. 59′. 35″.when the Nodes are in Quadrature with the Sun; and 5°. 17′. 20″. when the are in Syzygys.[26]

And from the Longitude and Latitude thus found, and the given Obliquity of the Ecliptick 23°. 29′. the Right Ascension and Declination of the Moon will be found. [27]

The Horizontal Parallax of the Moon, when she is in the Syzygys at a mean Distance from the Earth, I make to be

---

[26] See Ch. 9, VI, pp. 126–8.

[27] From this paragraph to the end of the text, Newton gives numbers that will be of practical importance in applying the theory, but that have no bearing on the theory itself.

57'. 30". and her Horary Motion 33'. 32". 32'". and her apparent Diameter 31'. 30". But in her Quadratures at a mean Distance from the Earth, I make the Horizontal Parallax of the Moon to be 56'. 40". her Horary Motion 32'. 12". 2'". and her apparent Diameter 31'. 3".

The Moon in an Octant to the Sun, and at a mean Distance, hath her Centre distant from the Centre of the Earth about $60\frac{2}{9}$. of the Earth's Semi-Diameters.

The Sun's Horizontal Parallax I make to be 10". and its apparent Diameter at a mean Distance from the Earth, I make 32'. 15".

The Atmosphere of the Earth, by dispersing and refracting the Sun's Light, casts a Shadow as if it were an Opake Body, at least to the height of 40 or 50 Geographical Miles (by Geographical Mile I mean the sixtieth Part of a Degree of a great Circle, on the Earth's Surface) This Shadow falling upon the Moon in a Lunar Eclipse, makes the Earth's Shadow be the larger or broader. And to each Mile of the Earth's Atmosphere is correspondent a Second in the Moon's Disk, so that the Semidiameter of the Earth's Shadow projected upon the Disk of the Moon is to be encreased about 50 Seconds: or which is all one, in a Lunar Eclipse, the Horizontal Parallax of the Moon is to be encreased in the Ratio of about 70 to 69.

Thus far the Theory of this Incomparable Mathematician. And if we had many Places of the Moon accurately observed, especially about her Quadratures, and these well compared with her Places at the same time calculated according to this Theory; it would then appear whether there yet remain any other sensible Equations, which when accounted for, might serve to improve and enlarge this Theory.

Dr. Greg Astr. Elem. Phys.
Geom. p. 336.
FINIS.

Part II

Commentary and Analysis

Chapter 1

Introduction

I. The Problem

Writing in 1975, Bernard Cohen posed the challenge:

> It would be most useful to have a careful analysis of Newton's attempts to produce a satisfactory lunar theory (in the 1690's), and the stages whereby he either partially or totally abandoned the program of deriving such a theory by mathematical methods applied to gravitational celestial mechanics (Cohen 1975, p. 80).

Cohen offered no comment upon either the accuracy of the theory—whether it was an improvement upon those available, or to what extent, if any, it was based upon a theory of gravitation. As Craig Waff commented in his review of Cohen's book:

> While I can sympathise with Cohen's reluctance to become involved in what would certainly be an extremely complex study, his failure to make even the slightest effort in this direction made it impossible for him to answer in any satisfying way a question which he constantly raises [i.e., that mentioned above]. (Waff 1977, p. 66).

Craig Waff commented upon the historical irony, that the brief 1702 essay *Theory of the Moon's Motion* was "probably the most obscure of Newton's publications, and yet it "appeared in print during the early eighteenth century more times than anything else which left the hand of Newton." His conclusion was that:

> Newton's 'rules' had been wholly or partially used by nearly a
> dozen astronomers or other interested individuals in order to
> construct lunar tables. (in Cohen 1975, p. 79)

Such a view differs radically from the more traditional opinion expressed by the astronomer Francis Baily, in his *Account of the Revd. John Flamsteed* (1837), to the effect that only a very few astronomers had ever attempted to apply it, and those that did had not found it much of an improvement upon existing sets of tables.

What strikes the modern reader about the text of *TMM*, apart from its obscurity, is the complete absence of any reference to a theory of gravitation. The *Principia* of 1687 dealt with motion under central forces as a one-body problem, and referred only relatively briefly to irregularities in lunar motion resulting from its motion as a three-body problem (Prop. 66 of Book 1, Prop. 32 of Book III). William Whiston gave the following fine eulogy to *TMM*, published in 1707 when he was occupying the Lucasian mathematics chair at Cambridge, as Newton's successor:

> The Moon, I say, which is a secondary planet, that hath in it such a complication of Motion, such intricacies and perplex'd Anomalies, that unto this very Day we are scarce able to bring it under Numbers, altho' it be so harrass'd (as it were) with Astronomical Researches. This hath been a knot well worthy of, and which requir'd the acutest Wit to untie. Nor wanted it such a one at length when the famous Sir Isaac Newton set himself to it; who hath this to glory in, That in the Compass of a few pages, he hath brought more light into this dark and intricate Business, than all the Volumes of the past ages had done. (Whiston 1728, pp. 95–6)

So finally, the Moon had met its match. Or had it? When Whiston came to explain how the Moon's position should be calculated, he said that, no doubt Mr Newton's theory was very excellent, however as no-one had yet reduced it to a form in which tables could be derived from it, he would give the rules as described by "the famous Mr Flamsteed." (Whiston 1728, p. 96) The abyss between theory and practice had not in fact been bridged.

No historian of science has acknowledged the validity of that judgement of Whiston, though he was in a fine position to assess the situation: that, in the year 1707, the procedure advocated by the Astronomer Royal was preferred to the Newtonian lunar rules, by the Lucasian mathematics professor, because the latter had not yet been unpacked, as it were. Was Flamsteed concerned to develop a lunar theory of his own? This view was somewhat indicated by the astronomer Francis Baily, who rescued Flamsteed's reputation from mere oblivion with his *Account* of 1835 (Baily 1837, p. 703). On the oft-told version of events, Flamsteed was allocated no other role than delaying or perhaps refusing to supply Newton with lunar data in the 1690s, thereby impeding the formation of the Newtonian lunar theory! That the Astronomer Royal may in some measure have fulfilled the mandate of the Monarch who appointed him, by achieving an improvement in the lunar rules, is seldom considered.

The brief 1702 opus *TMM* reaches into the future in two different ways: as a series of no less than seventeen reprints appearing in the first half of the eighteenth century (Cohen 1975, p. 6), and then secondly as the greatly expanded lunar section in Book Three of the second edition of the *Principia* of 1713. Later on we will comment on the apparent disappearance of Flamsteed's version of the theory, and its strange migration across the English Channel.

We may conveniently trace three steps in the development of Newton's lunar theory, for the period that concerns us. Firstly, in 1694–5 the extensive correspondence with Flamsteed recorded a keen collaboration, when the mathematician clearly believed he could encompass the irregularities of the Moon's motion by applying his theory of gravitation. Whiteside has well described how this noble enterprise became shipwrecked in the spring of 1695 upon the sheer intractability of the problem. Newton's decision to move to London and abandon his lecturing post at Cambridge, he suggested, may have been a consequence of his recognised failure with the lunar theory (*Mathematical Papers* 1976, VII, p. xxv). This phase ends as we shall see with Newton's despair, and his wrathfully prohibiting Flamsteed from mentioning the subject in public, after the latter had spent so much time in gathering the lunar observations required, on the grounds that nothing much was likely to come out of it.

Secondly, there was the *TMM*, published using Flamsteed's data but without the latter's knowledge or consent and despite two signed promises not to do such. As if in reaction against the failure of the first stage of the endeavour, no comment was made about a theory of gravitation. Thirdly, a decade later, there appear the mature Newtonian comments upon the three-body problem, which greatly impressed the cognoscenti. A review in the *Acta Eruditorum* (believed to be by Leibniz) commented on this section of the 1713 *Principia*:

> Indeed, the computation made of the lunar motions from their own causes, by using the theory of gravity, the phenomena being in accord, proves the divine force of intellect and the outstanding sagacity of the discoveror. (Cohen 1975, p. 41)

And Laplace said,

> Je n'hésite point à les regarder comme une des parties les plus profondes de cette admirable ouvrage. (Cohen 1975, p. 41)

A modern evaluation of the achievement should perhaps start from Owen Gingerich's claim, based on computer studies at the Harvard University astrophysics department, that little by way of increase in accuracy of ephemerides appeared as a result of the Newtonian revolution.

> Perhaps the most surprising result of our analysis is how little immediate and direct impact Newton's work had on the

computation of astronomical positions" (Gingerich & Welther
     1983, p. xi).
Was it, then, a theoretical affair? During the period which we are reviewing,
Paris became the main centre of ephemerides-production.

II. Accuracy of the "Theory"

Could Luna's erratic path across the night sky be used to ascertain longitude? Several notable disasters at sea stimulated astronomers to work
with greater zeal upon this nigh-impossible quest:

1691: seven British warships wrecked near Plymouth, mistaking the
Deadman for Berry head due to a misconception over longitude.

1694: Admiral Wheeler's fleet, ignorant of its position, sailed head-on
into Gibraltar and disaster.

1707: Sir Cloudsley Shovell's squadron of the Royal Navy ran onto rocks
off the Scilly Isles, with loss of four ships and nearly two thousand lives,
when they were believed to be in a safe position.

The last of these was due more to inadequacy in the maps used than longitude determination, as was shown by Commander Howse (1993, p. 47),
however it did much to arouse public opinion on the matter, and led to the
passing of the Longitude Act: in 1714, huge rewards were offered by
Parliament for anyone who could devise a method of locating the longitude
on a ship, to within one degree or less. "Finding the longitude" entered the
vernacular as meaning an impossible task which one despaired of ever
achieving. A life and death issue, it revolved around the most obscure equations. Britain was the only country that actually paid out rewards for the
achieving of this goal.

A Board of Longitude was established to investigate the claims, of which
both the Astronomer Royal and the Royal Society's President were mem-
bers. Its rewards began at £10,000 for anyone who could find the position at
sea to within sixty nautical miles, as meant predicting longitude to within a
degree. Finding the longitude within a degree meant that the Moon's longitude could be found to within two arcminutes (as the next section will
explain). The highest reward of £20,000 went for finding longitude to half a
degree, or within thirty nautical miles. In this context let us review the
bewildering spectrum of judgements that have been made over the
accuracy of *TMM*.

Within two arcminutes: This claim was brazenly made by David Gregory
in publishing the essay in his *Astronomicae Physicae et Geometricae Elementa*

of 1702, and no doubt stimulated its sales. A two-minute accuracy in lunar prediction would be sufficient to attain a one-degree accuracy in the estimation of longitude. Here, one cannot help recalling Flamsteed's characterisation of Gregory as a "closet astronomer" on the grounds that he made no outdoor observations. Rather surprisingly, this claim has recently been defended by Sir Alan Cook (Willmoth 1997, p. 186; Cook 1998, p. 372.)

Two to three arcminutes: this was Newton's own view as expressed in a 1705 edition, given in some corrections to the text which he inserted: two minutes in syzygies, three in quadratures. When Gregory republished *TMM* in the English translation of his book in 1715, he echoed this view. Thus, reprinting the essay 13 years later, Gregory hardly found cause to alter his original judgement of its accuracy. William Whiston made much the same claim in his published astronomy lectures of 1707. The astronomy professors of Oxford and Cambridge thus concurred in this formidable affirmation of *TMM*'s accuracy.

Two to five arcminutes: three decades later, Edmond Halley as Astronomer Royal affirmed that after himself preparing tables and ascertaining his calculation procedures, it was evident to him that:

> Sir Isaac had spared no Part of that Sagacity and Industry so peculiar to himself, in settling the epochs, and other Elements of the Lunar Astronomy: the result many times, for whole months together, rarely differing two Minutes of Motion from the Observations themselves. (Phil. Trans. 1732, p. 191)

Halley went on to say that, on occasions where the theory did err up to five minutes, this was probably the fault of the observer, i.e. Flamsteed, who had both supplied inaccurate data and failed to supply any in the third and fourth quarters of the lunar cycle. Halley was in a fine, indeed optimal, position to comment, though there is no reason to take seriously these slurs upon his predecessor. The latter's lunar positions achieved an accuracy of around half an arcminute (Kollerstrom and Yallop 1995 p. 242) and covered the entire lunar cycle.

Four to Seven minutes: this seems to have been Newton's estimate when appointed in 1714 to the Board of Longitude. Lunar methods he judged to be too inaccurate to determine "a Longitude within Two or Three Degrees." The French astronomer Bailly concluded in 1779 that the Newtonan theory was liable to err by up to five arcminutes.

Eight to nine minutes: this was the recent verdict of Curtis Wilson, editor of volume 2A of *The General History of Astronomy*, p. 267. He was merely echoing Flamsteed's verdict. The latter found, in the beginning of the year

1703, that *TMM* generated errors which were "frequently" of ±5 or 6 arcminutes, and that sometimes the errors rose to 8' or 9' in longitude at positions near to quadrature (ie, the half-Moon position). These things, he explained to his correpondent Mr Caswell, he determined using old data between the years 1675 and 1689, i.e. prior to the setting up of his great mural arc. Plainly he would not use data gathered since that date, as it had all been sent to Newton so that he could construct his theory. The astronomer was especially shocked by the errors in lunar latitude contained in *TMM*, which he said "were frequently 2, 3, or 4 minutes, which is intolerable." Examining lunar eclipse data, where one might expect smaller errors, on the grounds that astronomy had traditionally concerned itself only with the syzygy positions in the lunar orbit, for the prediction of eclipses, he again discerned errors of 5–6' in their positions (Baily 1837, pp. 213–4).

We will have more to say later concerning the protean flexibility of the estimates here represented. To place them in perspective let us cite some findings of Gingerich reported in 1983: that *La Connoissance Des Temps*, the ephemeris then produced yearly by Cassini from the Paris Observatory, did over the years 1695–1701 frequently display errors in its lunar positions of 20–30 minutes of arc; and secondly, that the much-desired predictive accuracy to two minutes of arc was not attained by any ephemeris prior to the British Nautical Almanack commencing in 1765 (Gingerich & Welther 1983, Fig. 14 and p. xxi).

In using modern programs to resolve this issue two problems arise. Firstly, any modern estimate will take the form of a mean and standard deviation. We might relate this to a historically-expressed limit by doubling it, for example. Two standard deviations enclose 95% of a distribution. Secondly, a computer-reconstructed version of *TMM* might not have taken account of errors involved in interpolating the required values from tables. It might thereby give what was in practice too small an error-value. Later chapters will have to deal with these issues.

III. The Two Clock Method

The latitude of a ship could readily be found from the altitude of the pole star. There was no problem in this. In contrast, the finding of longitude had become the most urgent scientific problem of the age, being the reason why Charles II established the Observatory at Greenwich. This need stimulated the quest for versions of *universal time*. Local time varies around the globe by one hour per fifteen degrees of longitude, and so, if one had two clocks, one on local time and the other on universal time, longitude would come from the time difference between them. At sea, a clock is set to local time by using sunrise and sunset, where noon falls midway between them. If the Moon's 27.3 day cycle against the stars could be determined, then it would enable one to read universal time, as a clock whose hand revolved once in twenty-seven days. That was the beckoning dream, the impossible hope, the mirage on the horizon...

The Earth spins in space twenty-seven times faster than its Moon revolves around it. That ratio gives *the error-multiplication factor* inherent in the method. If the lunar longitude is out by one arcminute, then the terrestrial longitude thereby obtained will err by twenty-seven arcminutes, ignoring other sources of error. An error in lunar longitude translated into an error in the Universal time found; this time-error becomes expressed as a fraction of the Earth's daily rotation, ie an error in longitude. The Board of Longitude set up a goal, as implied that lunar celestial longitude would be predicted within somewhat less than two arc minutes. This would find the longitude within a useful - though not a safe - range. Without that one would be, as the saying went, "at sea." To find longitude within a degree meant predicting the Moon's position within $1/27.3$ degrees = 2.1 minutes. For comparison, the English Channel spans just over half a degree of longitude.

In the seventeenth century, not the least source of error in using the method was the absence of any sound notion of mean time. The observations were made in local apparent time and then had to be converted to "equall time" (i.e., local mean time). Only then could a comparison with universal time, from the lunar sidereal orbit, be accomplished. Tables for this conversion were (I found) wildly inaccurate: for example, the "Table of the Aequation of Civill Dayes" given in Wing's *Harmonicon Coeleste* of 1651 had an average error of five minutes. Such a time-error would give a two or three minute error in lunar longitude. I also checked the equations of time given by Streete (1664) in several of his worked examples, plus a column of such figures given in the French annual ephemeris *La Connoisance des Temps* of 1686: which showed errors usually around 4–5 minutes.

The first reliable Equation of Time was published by Flamsteed in 1673, in a postscript to the *Opera* of Horrocks, (Flamsteed 1673, pp. 441–464), then a later more accurate table (with at most 10–12 seconds or error) of his was published by Whiston in 1707. Historians generally give little credit to Flamsteed for establishing Greenwich mean time, by discerning that the Earth's uniform sidereal rotation throughout the year should be its basis, though the French historian Bailly did admit that he had "restored" the equation of time (Bailly 1779, p. 269). A substantial improvement in lunar longitude determination thereby came about. *TMM* simply presupposed that mean time was used. Flamsteed ascertained Earth's uniform rotation rate, by timing the daily transit of Sirius, as provided the basis for his concept of mean time.

How accurate was the method in practice? An example comes from an entry in the diary of Edmond Halley, when he landed his frigate off the coast of Brazil in the year 1699. He was returning from his courageous antarctic voyage (which was more dangerous than he seems to have realised, navigating past huge icefloes with their colonies of penguins), and wanted to find out his longitude. He and the crew of his "Pink" (a type of Dutch sailing vessel), the *Paramore*, found themselves near the town of Paraiba.

Halley first of all set up his telescope to view Jupiter, because his tables predicted an occultation of one of the Jovian satellites, on the night of February 25th. That appears, let us note, as having then been his preferred method (The four Jupiter satellites obeyed Kepler's three laws exactly, and so their appearances and disappearances gave a convenient measure of universal time). In this he found himself frustrated, because clouds obscured his view. The Jupiter-moon method being unusable, it happened that the Moon was passing by a first-magnitude zodiac star Antares, whereby an "appulse" could be observed. An "appulse" meant the time of nearest approach of two heavenly bodies. Halley noted the time of this event to the nearest second, and the lunar altitude when it occurred, and from these wrote:

I conclude the longitude of this Coast full 36° to the Westward
of London.

Halley was within almost one degree of the correct longitude, which is quite impressive. The longitude of Paraiba is 34° 52' West (Thrower 1981, p. 103; Kollerstrom 1990, p. 7). The ephemeris he used was probably the French *La Connoissance des Temps*.

When Newton sat on the Board of Longitude, set up in 1714, he there expressed the view that the lunar method only worked "within two or three degrees" (Westfall 1980, p. 835). This would have been equivalent to finding lunar celestial longitude within 4–7 arcminutes, which sounds fairly reasonable. This suggests that Halley obtaining longitude within almost one degree, two decades earlier, was something of a fluke. Or, perhaps, the people of Paraiba did know their longitude.

The lunar method was first used with success at sea in 1753, by Nicholas-Louis de Lacaille in an Atlantic crossing (Howse 1993, p. 4). An account of the lunar method of finding longitude given by Howse emphasised the success which the method eventually enjoyed, from the latter half of the eighteenth century: "The heyday of lunars was probably from about 1780 to 1840" (Howse 1996, p. 159). Marine chronometers did indeed become available from the mid-eighteenth century onwards. However, they remained prohibitively expensive for most vessels, so that the Greenwich *Nautical Almanac* published annually from 1767 offered the preferred method of finding longitude at sea. It is often averred that the invention of the chronometer rendered the lunar-distance method obsolete; in fact, ten thousand copies of the *Nautical Almanac's* 1781 edition were sold immediately upon publication (Sadler 1976, p. 120). Chronometers "never came into popular use at sea until the middle of the nineteenth-century" (Cotter p. 243).

Merchant vessels using the lunar method came to adopt the Greenwich meridian for their reference, as the lunar positions of the *Nautical Almanac* were for Greenwich time. The ephemerides of other nations also reproduced these positions, as being the best available (Sadler 1976, p. 120). Thus the work of Newton and Flamsteed did in the end bear fruit, a century or so later, with the locus of their endeavours eventually becoming accepted globally as the zero meridian of longitude.

The British *Nautical Almanac* began publication in 1767, and gave lunar celestial longitude at three-hourly intervals in *apparent* Greenwich time. The Gingerich & Welther graphs of its error patterns over 1779-1787 show them as generally between one and two minutes of arc. Then, "By 1800 the accuracy of the best almanacs was…better than a minute of arc" (Gingerich and Welther 1983 p. xi). This is compatible with a prevalent view that lunar tables of the 1760s enabled sailors consistently to find their longitude "within 1°" (Sadler 1976, p. 117; and Kelly 1991, p. 243), for which a two-minute error in tables of lunar longitude might have been tolerable. In contrast with these views, I checked the first page of the *Nautical Almanac* for January 1767 for the first twenty lunar meridian transits, and found a mean error of 16" ±17".

I found a several-fold smaller error pattern in this data than had Gingerich and Welther (1983, p. xxi): the Yallop et al. program for Equation of Time (Hughes, Yallop, and Hohenkerk 1989) was used, to convert from apparent to mean time, plus an I.L.E. program for lunar longitude. For example, at noon on January 7th, 1767 the Almanac gave lunar longitude as 18° 59' 25" Aries. Subtracting the Equation of Time, 6 minutes and 50 seconds, gave a GMT value of 11am 53 minutes and 10 seconds, where

mean time = apparent time – equation of Time,

for which the true (modern computed) position was 18° 59' 16". The net error-value (i.e., historic less the computed value) was 9″. These results well accord with what was believed at the time about the tables, and help us to appreciate the extent to which the lunar-longitude method did in the end *succeed*. This case-study underscores the vital importance of having computer programs more accurate than the historical positions to be evaluated.

IV. Differing Approaches to the Problem

In the latter half of the seventeenth century, we may separate out several approaches to predicting the Moon's position. They intertwined, but were logically distinct.

1. Use of a Model: the Method of Horrocks

As a north-countryman, Flamsteed was proud of having made public and improved the technique invented by the young Jeremiah Horrocks in the 1630s. Horrocks invented a kinematic model, wheels within wheels, like some English Heath-Robinson version of the epicycles so recently banished by Kepler: but, it worked. His theory involved the rocking motion in the apse line of the Moon (once per six and a half months) conjoined with a large fluctuation in its eccentricity over the same period. The version published in 1673 was the Horrocksian method improved by Flamsteed. He thereby became eligible for the newly-created job of Astronomer Royal. Perhaps for the first time in history, a professional astronomer would not cast horoscopes, but would instead seek out a method of finding the longitude.

2. Empirical: Using the Saros

The newly-wed Edmond Halley commenced in the year 1682 taking lunar longitude readings with a view to tracing a whole Saros cycle of 223 lunar months. It happens that all the principal irregularities in the Moon's motion repeat through this period in a precise and cyclic manner. Where he got this fine idea from is unclear, as the *Principia* does not mention it. The notion appears as distinctively belonging to Halley. He believed that a continuous sequence of observations over such a period was the best approach.

Sir Alan Cook has found that Halley was then living in what is now the Hackney area of North london, although no plaque marks the spot. Halley's program was terminated abruptly by his Father's murder, the body washed up on the banks of the Rochester river in Kent in April of 1684 (Cook 1998,

p. 432). Halley came to fulfil his plan six decades later, as does indeed show streadfastness of purpose. He followed a complete Saros of meridian transits taking them every third day or so at Greenwich, though his successors did not deem his observations of much value.

3. Mathematical: The Theory of Gravity

> For I find this theory so very intricate, and the theory of gravity so necessary to it, that I am satisfied it will never be perfected but by somebody who understands the theory of gravity as well, or better than I do.

Newton wrote these words to Flamsteed on February 16, 1694. They were to be vindicated by the mighty labours of Clairaut, Lagrange and Laplace in the next century, using the Leibnizian calculus. But in that period, Newton was faced by abject failure: he later wrote wrathfully to Flamsteed on hearing that the latter proposed to make public the fact that he had supplied 150 lunar positions for the Cambridge mathematican, by way of explaining what he had been up to:

> I was concerned to be publicly brought upon the stage about what, perhaps, will never be fitted for the public, and thereby the world put into an expectation of what, perhaps, they are never like to have. (*Corr.* IV, January 6, 1699).

That was his last known comment upon his endeavour with the lunar theory prior to *TMM*'s composition, which is curious.

V. The Apse Line Motion

What is called the "apse line" joins together the Moon's apogee and perigee positions. If we view the lunar orbit as an ellipse, with Earth at one focus, this line forms its long axis. It revolves against the stars once per nine years. This definition needs to be modified by saying that such a line joins together the *mean* apogee and perigee positions. The line is a mathematical abstraction, because the true apogee and perigee positions can deviate widely from it. Perigee swings nearly 30° to and fro, twice a year The apogee and perigee positions oscillate in rather different ways, and do not remain opposite each other in the sky.

One may wish to represent these half-yearly oscillations by viewing the apse line itself as rocking back and forth. That is how the matter was viewed by Newton and his colleagues. A theoretical model represented the phenomena. A crucial concept used to describe its working was "equated": a mean apse moved with uniform "motion," i.e., angular velocity, around the ecliptic, and the apse was then "equated" by adding the oscillating motion.

Horrocks envisaged this secondary motion in 1638. Curtis Wilson has shown how Horrocks reached his new model from re-analysing the rather unsatisfactory lunar theory of Kepler (Wilson 1987 p. 78). Only later on, in the 1640s, did the North-country astronomers Gascoigne and Crabtree provide confirmatory evidence by studying the Moon's varying apparent diameter (Chapman 1982, pp. 19–21). The new telescope technology they were using with eyepiece micrometers enabled them to measure the changing apparent lunar diameter. Before that, no one could well discern the motion of the apse line. A letter of Flamsteed's printed in the Royal Society's *Phil. Trans.* of 1675 described how he came to adopt the Horrocksian system: it was only after "many curious and careful measures of the *Moons diameters*" that he came to realise that no other theory could account for the phenomena *(Phil. Trans.* 1675, pp. 368–370). An inequality called "evection" was related to this effect, as being the largest of the lunar inequalities, first given that name by Ismaël Boulliau in 1645.

Here is how Newton's *System of the World* described the matter:

> By the same theory of gravity, the Moon's apogee goes forwards at the greatest rate when it is either in conjunction with or in opposition to the sun, but in its quadratures with the sun it goes backwards; and the eccentricity comes, in the former case, to its greatest quantity; in the latter, to its least. (*PNPM* p. 660)

This gave a "semiannual equation of the apogee," of amplitude 12° 18' "as nearly as I could determine from the phenomena."

This Horroxian theory was in use until the mid-eighteenth century. Continental mathematicians then reanalysed the situation and seemed not to need this concept. To quote the French theoretical astronomer Clairaut given in 1748, in a letter to the Astronomer Royal Bradley (translated from the French):

> At bottom, my solution of the lunar theory is quite different from that of Mr Newton: I do not find as he did the variations in eccentricity or inequalities in the movement of the apogee. (Gaythorpe 1956, p. 136)

That was the Horrocksian theory which Clairaut was rejecting. We may take a partisan view here, and affirm that the lunar apogee does indeed have such inequalities, as Horrocks ascribed to it, more or less. It is rather more difficult to see how its eccentricity varies as Horrocks described, as modern theory lacks anything resembling the ±21% variation in the eccentricity function, that *TMM* utilised. This matter is further discussed in the next chapter.

If one turns to a nineteenth-century account of these things, say Stevenson's *Newton's Lunar Theory Exhibited Analytically*, (1834) then what there majestically unfolds as "Newton's Lunar Theory" has no trace of that double motion of the apse line: it has merely a single rotation in nine years. The whole thing much resembles Clairaut's lunar theory, and that of Horrocks is nowhere to be seen. Clairaut's view, to quote further from his letter to Bradley, was: "les différentes espèces de termes qui sont dans mon equation pourront bien faire le même effet que les variations dans l'excentricité et dans le mouvement de l'apogee." Stevenson's 1834 version thus apears as a mythologised version of the "Newtonian theory". In a preface the author assures us he has merely translated the theory "from the hieroglyphics of geometry" into the workaday language of algebra.

Herein lies the nub of the problem. The hieroglyphics of the geometric-kinematic forms in which seventeenth century lunar theory expressed itself may seem as remote from modern comprehension as an arcane alchemic sigil to a modern chemist. It will require quite an effort on our part to enter into the meaning of these old diagrams, from a period *before* trigonometric functions were used to describe the time-dependent variables of astronomy.

Whiteside's 1976 tercentenary essay marks the beginning of a realistic assessment of *TMM*. Entitled "From High Hope to Disenchantment," it has been accepted by more recent scholars, pre-eminently Curtis Wilson, whose fine achievement in the *General History of Astronomy* (vol. 2A) includes an evaluation of *TMM*. It concludes that Newton's adoption of Horrocks theory was a historical mistake, which prevented his making further progress. That seems a rather pessimistic view. Also it may not adequately assess the extent to which Horrocks' theory was *true*. The young Horrocks has after all been viewed as initiating the tradition of British astronomy (Chapman 1982 p. 8).

VI. Enlightenment Reception of TMM

The French astronomer Bailly struck a sceptical note over *TMM* not found amongst English historians: "but he [Newton] has often spoken in the manner of the prophets, who speak of that which one cannot see." (Bailly 1782, III, p. 150). Despite this, Craig Waff and Curtis Wilson both affirmed that Newton's lunar theory was applied to the construction of lunar ephemerides in the first half of the eighteenth century. Wilson described how *TMM*'s procedures "were incorporated in the tables of Charles Leadbetter's *Uranoscopia* (1735)." (Wilson and Taton 1989, p. 269). Baily had said the same in 1837, affirming that Leadbetter had given "a more perfect adoption of Gregory's *Newtonian rules* [Baily's term for *TMM*] reduced to a tabular form." (Baily 1837, p. 709). As Francis Baily pointed out, however, in 1742 Leadbetter brought out a new set of lunar tables, "without any allusion to Newton's labours."

As for the work published by Leadbetter in 1735, its frontispiece merely states that the book will give the "Flamsteedian method of Computing times of Eclipses." Chapter nine of his *Uranoscopia* is entitled, "to calculate the true Place of the Moon more exactly than was ever yet done"; however, this contained no allusion to *TMM*. Then the last chapter, introducing its tables, again contains no Newtonian allusions.

Leadbetter's allusion to *TMM* was ironical in tone. Discussing the work of a rival, Leadbetter advised his readers,

> Another tells us, that his Calculations are from Sir Isaac Newton's Theory of the Moon; and therefore nobody must question the truth of them. Indeed, if it were so, not any one living would dare to question them. But I deny the assertion; and can prove, that his calculation is not from Sir Isaac Newton's theory.

Thus, kudos appears to have been available to any almanack claiming to be based upon *TMM*, and some rivalry is here evident. Leadbetter compared several published predictions for an eclipse, and claimed to have made his own prediction "from new Tables, founded upon Sir Isaac Newton's Theory of the Moon," which gave the most accurate eclipse time. Does that amount to a claim that the tables of his book had been derived from *TMM*? If so, the claim was made in an equivocal fashion. No such claim was made either in the two relevant chapters, or on the frontispiece. Elsewhere, Leadbetter alludes to a rival: "Tycho Wing, in Coley's Almanack, which he says is from Sir Isaac Newton's Theory of the Moon; but this is a mistake, because it is so vastly wide of the truth, that it will not bear the test." Leadbetter was claiming to have fathomed *TMM*, without going so far as to say that his own tables were based upon it.

Chapter 2

Lunar Inequalities and Apse Motion

I. Stages of Development

The lines of Halley's ode,

> *At last we learn wherefore the silver moon*
> *Once seemed to travel with unequal steps,*
> *As if she scorned to suit her pace to numbers—*
> *Till now made clear to no astronomer,*

were composed for the 1687 edition of *PNPM*, yet hardly applied to it. Rather, they expressed what Halley as an astronomer hoped he would find in it. They did however apply to the 1713 edition. The first edition of the *Principia*, as Whiteside (1976) emphasized in his 1975 tercentenary address over the founding of the Royal Greenwich Observatory, dealt most successfully with lunar motion as uniform and regular, i.e. as a one-body problem concerning the motion of a body around an immovable force-centre. In the Second Edition, the position of the barycentre (Earth-Moon centre of gravity) was estimated, enabling two-body computations to be for the first time performed.

That accomplishment of 1687 did not assist the construction of lunar ephemerides. The first step in this direction came following a visit by Newton to Flamsteed at Greenwich in November of 1694, when he was shown a table comparing observed and theoretically-derived lunar longitudes over a series of meridian-transits. The theoretical positions had been derived from Flamsteed's Horrocksian method as published in 1681, and were compared with lunar centre positions for the same times, obtained from his observations of lunar limb transits. A column of differences showed the errors in computed longitudes, as averaging around eight minutes of arc. On the whole, Flamsteed's determinations were within half a minute of error, though cited to arcseconds.

Newton was able to borrow these tabulated data, and then came to request altogether just over two hundred lunar positions from Flamsteed: which he was sent—contrary to centuries of calumny about the latter refusing to part with his data—in the months following (Kollerstrom & Yallop 1995 p. 238). No mathematician hitherto had had so many lunar positions of such accuracy. In the early months of 1695, Newton's letters to

Flamsteed display a keen enthusiasm for the subject, and belief that his theory of gravity should be able to encompass the problem. After all, the rest of the Universe was obeying it.

This phase of Newton's lunar endeavour terminated rather abruptly in 1695, shortly prior to his moving to London and becoming Warden of the Mint. Optimism gave way to bitterness, and what had been a friendly and respectful correspondence since 1672 (when Flamsteed wrote to Newton over the latter's new colour theory) was replaced thenceforth by distrust, at least on the astronomer's part. Flamsteed was not permitted to claim credit for his labours in producing the lunar data, and the 1702 treatise appeared without his knowledge or consent.

TMM, written by the Master of the Mint, surveyed the periods and inequalities of lunar motion, and described a kinematic model, basically that of Horrocks. It thus represents a diametric antithesis to the *Principia's* endeavour of 1687. The latter was a work of theory, of zero practical utility as far as lunar prediction was concerned. The former contained no theory as is nowadays understood (despite its title, conferred it is supposed by David Gregory), and gives no hint that its author had developed an inverse square law of gravity. It is as if the hope expressed in early 1695 had been extinguished, in that no theory was present, and its author had regressed to a kinematic approach, with the old epicycles and deferents still there. The frequent reprinting of *TMM* through the first half of the eighteenth century indicates that it was highly esteemed as of practical utility.

II. The Horroxian Ellipse

Jeremiah Horrocks was, as Bernard Cohen observed, "the first thorough-going Keplerian in England" (Cohen 1975, p. 11). His new lunar theory applied the Keplerian concept of elliptical motion to the lunar orbit, repre-senting the monthly motion between apogee and perigee in such terms. This might sound straightforward enough; however, there was an intractable prob-lem associated with such a step. What Brahe and Kepler had termed "the Variation" was the deformation of a *circular* lunar orbit resulting from the Sun's pull, such that the Sun-Earth axis defined its minor axis. The orbit there-by became flattened into an elliptical shape, with Earth at the *centre* and elon-gated at the quarter-positions. This 'variation-ellipse' revolved yearly against the stars, its minor axis always facing the Sun. Newton's Variation-ellipse had an eccentricity considerably larger than that of the apogee-perigee ellipse.

Could such a model for the Variation be conjoined with a different ellipse for the apogee-perigee motion? As Curtis Wilson observed, "The Variational orbit and the Horroxian ellipse implied two disparate pictures of the lunar orbit" (Wilson 1999, p. 24). The problem was inherent in the Horroxian model. In the year 1642, William Crabtree wrote to Gascoigne giving the first account of the new, Horrocksian theory—in seven steps— and his final paragraph seems to express his perplexity:

> In this way, as I described above, I find the place of the Moon investigated in all his last calculations; the precepts for calcu- lating the place of the Moon in its orbit are all based on its elliptical path, except for the 7th, concerning the Variation; how this can be applied demonstratively to this theory, escapes me; I pray you, give your thought to the matter. (Horrox 1673, p. 472)

The last of his seven steps concerned the Variation. Crabtree was far from being the last to become perplexed over the issue. Newton came to the view, doubtless on account of this very problem, that: "the eccentric orbit in which the Moon really revolves is not an ellipse but an oval of another kind"(Whiteside, 1974, VI, p. 519). The dynamical problems here involved hardly impinge upon the present study, yet we are concerned with the manner in which the Kepler-ellipse concept was applied to the lunar orbit.

The Newtonian lunar "theory" has main components, which we may view as additive: (1) his "equation of the centre," a variant of Seth Ward's "empty focus" method of approximating to Kepler's second law; (2) the Horrocksian oscillation of the apse line, with its concurrent oscillation in the eccentricity of the lunar orbit; and (3), six extra lunar "equations" added to these, which were original. His method of computing the Horrocksian oscil- lation eventually (1713) came to use an approach of Edmond Halley, where- by the lunar ellipse had its center on an epicycle which revolved twice yearly around a point near the Earth, modified by a second small epicycle revolving around its perimeter.

III. The Rocking Apse

> And those inequalities…generate the principle which I call the semiannual equation of the apogee; and this semiannual equa- tion in its greatest quantity comes to about 12° 18′, as nearly as I could determine from the phenomena. (*PNPM* p. 660–1)

Over a period of one year and forty-five days, the Sun-Earth axis aligns twice with the apse line. Let us call this period the Horroxian year, as there is no current astronomical term for it. The latinised form of Horrocks' name will here be used, for such astronomical terms as pertain to his theory. Over half that period, so Jeremiah Horrocks in 1638 affirmed, the apse line had an oscillation of 12° amplitude. It swung dramatically back and forward twice, in addition to its yearly mean motion of 40°. Newton referred to this as "the semiannual equation" (Scholium of Prop. 35, *PNPM*, p. 661), by which he meant that its period was half a year. More precisely, its period is 206 days. The duration of the "Horroxian year" comes from the equation,

$$1/365.24 \quad - \quad 1/3232.6 \text{ days} \quad = \quad 1/411.7 \text{ days.}$$

$$\text{year} \qquad \text{apse rotation (8.6 yrs)} \quad \text{Horroxian year}$$

This equation has two cycles per Horroxian year. By plotting the longitudes of apogee and perigee positions each month, we may inspect this claim (Figure 2.1).

In *TMM* of 1702, Newton gave the "greatest Equation of the Apogee" as 12° 15' 4". This meant, its greatest deviation from a mean position. In *PNPM* of 1713 this increased to 12° 18'. An observer of the Moon's apogee, which is its position in the sky each month when it appears smallest, would perceive an oscillation in its ecliptic longitude of around two or three degrees only, twice each year, not twelve degrees. Thus, such an oscillation of the apogee is not to be found in the heavens.

The perigee position in contrast oscillates more vigorously, moving back and forth with an almost twenty degree amplitude. If we take an average of these two oscillations, only then does the figure of twelve degrees emerge. Thus the concept of an apse line is an approximation, since perigee diverges greatly from such an axis. The apogee and perigee positions have distinct motions. There is a contrast here with the nodal axis, where the two nodes appear as being closely opposite in the sky, and have a more uniform motion, so it makes sense to visualise a nodal axis between them. To quote from a modern work:

> The oscillations [of the apsides] do not take place simultaneously, but alternately, so that the apsides are not always directly opposite one another in the zodiac, but are continually falling behind and overtaking these positions. The retrograde motion of the perigee (roughly 40°) is very much larger than that of the apogee (roughly 2° to 3°), meaning that the former moves much more quickly than the latter against the fixed star

background...the perigee can regress by more than 30° in a
single month, whereas the apogee moves for up to four
months within a field of only about 3°.' (Schultz 1986, p. 91)
The graph (see over) illustrates these motions, measured in ecliptic longitude, of the apogee and perigee positions in modern times. I contructed it
using the times for these monthly events as given in Meeus' *Astronomical*

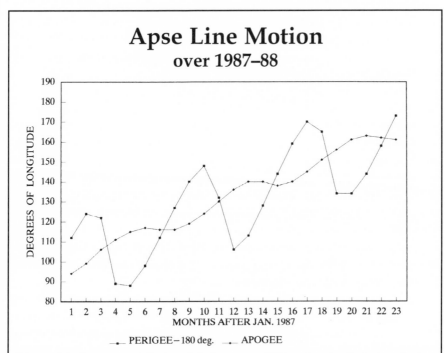

Figure 2.1: Apogee and Perigee Motions on a 180° scale of ecliptic longitude,
illustrating their coincidence on an "apse line," and the greater motion of perigee
as compared to apogee. Apogee and perigee positions have been reconstructed
from times supplied by Meeus (1983).Apogee positions are two weeks later than
those of perigee. 180° has been subtracted from the perigee positions to align
them with apogee.

Tables (Meeus 1983), and computed positions therefrom. The graph shows
the two cycles, each thirteen months, on top of a mean motion of the two
positions of 40° per annum.

A model approximates to reality. In this case, the Horroxian model took
the apogee and perigee motions as having a mirror-symmetry which they
lack in reality. The model was not primarily concerned to account for lunar

distance - reflected in its apparent size in the sky - but to predict longitude, so this may not have been too much of a disadvantage. The unwary reader of *PNPM* could here be misled on two counts: the inequality was not of the apogee as stated, but of the apse line; and it was only approximately half-yearly ("semiannual"). Also, because this function is discontinuous - there is only one perigee position per month, it has no existence in between these times - a degree of accuracy quoted to seconds is questionable.

We can now accept Newton's account as given in *PNPM*, if we just substitute the word "perigee" instead of "apogee":

> The moon's apogee goes forwards at the greatest rate when it
> is either in conjunction with or in opposition to the sun, but in
> its quadratures with the sun it goes backwards (p. 660).

In Figure 2.1, the maximal forward motion of perigee corresponds to solar alignment with the apse line (i.e., conjunct or in opposition to apogee), whereas its furthest retrograde position occurs at the quadratures. The converse applies for apogee.

Historians of astronomy tend only to discuss the mean apogee motion, of 3° per lunar month, and the historical problem of accounting for this motion by a gravity theory. They seldom acknowledge that the apse line really does have this rather interesting secondary motion, discussing the Horroxian model as if it were merely a reformulation of the antique concept of "evection." Rather, this motion was a fine British discovery by the young North-country clergyman Horrocks, and it formed the core of what Newton recognised as the best lunar model available in the seventeenth century. Corollary 7 to *PNPM*'s Proposition 66 of Book I (*PNPM* pp. 283–4) claims to deduce this oscillating motion from the theory of gravity.

IV A Varying Eccentricity

For Hipparchus, eccentricity meant Earth's distance from the centre of a circle, to which the lunar orbit approximated, the radius of that circle being taken as unity. With the entry of Keplerian ellipses, eccentricity still meant Earth's displacement from such a centre, but the circle circumscribed the orbit ellipse, i.e., its diameter equalled the long axis.

The first recognisably modern definition that I found for eccentricity was given by William Whiston, who defined it as the distance from the centre of an ellipse to its focus (Whiston 1728, p. 91). His textbook had a section, "To Determine the Earth's Eccentricity," which explained how eccentricity was "to be reckoned from Focus to Center." This was around 1703, for his

Lucasian lectures at Cambridge. Astronomers derived it, he explained, from comparing the angular velocities at apogee and perigee, since at these points the minimum and maximum values respectively were reached. By Kepler's second law, he explained, the relative lunar distances should be inversely related to these angular velocities. This same definition appears in a glossary of astronomical terms given by the eighteenth-century astronomer Leadbetter: "Eccentricity is the distance between the center of the ellipse and the focus" (Leadbetter 1742, Vol. I p. 19). This definition assumes the long axis of the ellipse (or, the radius of the circumscribing circle) is unity, whereas in practice it would be some large figure, one million in the case of *TMM*. Thereby the eccentricity is an integer value, say in thousands.

A simulation was performed of the manner in which astronomers of the period perceived lunar eccentricity to vary. If apogee and perigee distances are A and P, then it can readily be shown that eccentricity is given by

$$(A - P)/(A + P).$$

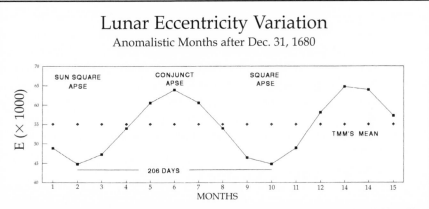

Figure 2.2: Monthly values of a simulated eccentricity value, (A-P)/(A+P), where A and P are mean apogee and perigee distances, for the months of 1680 and 1681. Its mean value was 0.0543.

The computer was set to generate lunar distances at successive mean apogee and perigee positions, and thereby generate estimates of eccentricity per anomalistic month. Figure 2.2 shows how this monthly function varied over a period of one year.

The figure shows lunar eccentricity peaking as Sun and Earth align with the lunar apse, in accord with the Horroxian theory. The amplitude of its oscillation appeared as a little less than that specified by *TMM*, as should be ±21% of the mean. William Whiston experienced some difficulty in visualising this rather flexible model:

> That the Inequality of the Eccentricity of the Lunar Orbit, which is greatest when the Line of the *Apsides* falls in the Conjunction or Opposition, and is then one and a half of what it is in the Quadratures; which consequently renders the Ellipsis perpetually mutable, sometimes coming nearer to a Circle, sometimes a great deal more remote from it, so as not to be reduc'd to any certain Species, and which is scarcely to be accurately defined…" (Whiston 1728, p. 130)

For comparison, the long axis of the lunar ellipse ($A + P$) in the above simulation was found to vary by only 1%. This well accords with Whiteside's account of the Horrocksian theory, as involving "a Keplerian ellipse (ever with the earth at its primary focus) whose major axis remains unchanged in length during its periodic distortions." (Whiteside 1974, VI, p. 509).

Since the lunar theory of Leonhard Euler in the mid-eighteenth century, eccentricity has been treated as a fixed, unvarying function. Modern textbooks do not generally give a value for this fluctuation, treating it as being invariant. Gaythorpe (1925 p. 860) explained how the modern evection terms were mathematically equivalent to Horrocks's varying eccentricity. An exception here is the Spencer Jones textbook, which cited a value oscillating by the Newtonian 21% (Spencer Jones 1961, p. 118).

The fluctuation in eccentricity every 206 days has the same period as that of the apse line's oscillation, of two cycles per Horroxian year. However, the two are out of phase: as the Sun aligns with the mean apse, the eccentricity equation maximises, while the perigee has then reached its mean position, i.e. its "second equation" is zero while its rate of change in ecliptic longitude is maximal. Chapter Seven will describe how Horrocks's model generated these two motions.

TMM added a new second step of the lunar equating process, fluctuating in phase with the apse line's conjunction with the Sun. It was again a second harmonic (i.e., a 2Θ function), zero when the Sun was conjunct the apse line or at quadrature to it, and maximal at the octant positions. Newton also called this equation "semiannual," giving it a mean magnitude of 3' 45" (*TMM*, p. 15 (p. 8 above); *PNPM*, p. 660). His extra node equation likewise varied in tune to the Sun's alignment with the nodal axis, which he called the "second semestrial equation." Again this maximised at octants, but had its period a month shorter than the Horroxian year, as the nodes revolve in the opposite direction (*TMM*, p. 17 (p. 12 above); *PNPM*, p. 660).

Chapter 3

Some Perspectives on TMM

I. From Minutes to Seconds

TMM cited its longitudes to seconds of arc. Did astronomers of that time really enjoy such accuracy? Curtis Wilson has ascertained that the ephemeris of Thomas Streete was the most exact of any available in the seventeenth-century (*GHA*, p. 180). Flamsteed wrote in 1669 that "I esteem Mr Streete's numbers the exactest of any extant" (*GHA*, p. 179). To give some idea of his data accuracy, let us consider a solar eclipse which Streete cited in his *Astronomia Carolina* of 1661. This was visible in London, on May 22nd 1639. Streete gave the "Apparent Time" of its end as six hours, 10 minutes, 27 seconds, from which he derived what he called its "Equal time" of six hours, zero minutes, 27 seconds. That is to say, Streete's "equation of time" was ten minutes, by subtracting which he converted to mean local time.

Eclipses were times when the Moon's position could be ascertained very exactly, by comparing with the Sun's position. Here we do in fact find arc-second accuracy; however, his "equation of time" was way out, erring by seven minutes! As we saw earlier (Ch. 1, III), astronomers of this period had only a hazy grasp of this notion. Far from being ten minutes, its correct value then, as ascertained using a modern program (Hughes, Yallop, and Hohenkerk 1989), I found to be less than three minutes. Streete presents his celestial longitudes as having been derived from computation, but his accuracy can hardly be taken seriously with so large a time-correction error.

For us to obtain the relevant GMT value, the equation of time and an adjustment for London's longitude (Appendix I) are subtracted from the given Local Apparent Time (LAT). As the computer shows, his longitudes for that moment were within arcseconds.

Conclusion of the solar eclipse of May 22, 1639

Streete's positions	*Modern-computed*	*differences*
Sun: 10° 49′ 28″ Gemini	10° 49′ 51″ Gemini	- 23″
Moon: 12° 01′ 01″ Gemini	12° 01′ 18″ Gemini	- 17″

LAT of 18hrs, 10min, 27sec.⇒ GMT 18hrs 7min 10sec, for eqn time 2min 57 sec.

For readings taken in 1639, they are quite impressive.

Moving on to the 1690s, research conducted by the present writer in collaboration with Bernard Yallop at the RGO has indicated that the stellar observations tabulated in Flamsteed's *Historia* tended to be within five seconds of arc or so accuracy. For example, on March 8th 1695, the star Aldebaran was cited as having a zenith distance of 15° 51' 24", in the *Historia Coelestis'* Volume II. Subtracting that from 51° 28' 10", which was Flamsteed's value for the latitude of Greenwich, then subtracting out the appropriate value for refraction corresponding to that altitude as given by modern tables, gave 35° 36' 18" declination. The computer determines the correct declination for Aldebaran at that time as having been 35° 36' 15", a difference of three seconds. Likewise for the star Spica on April 12th, 1698, the error was 5 seconds of arc. These vertical star readings are relatively simple to compare: there is no parallax correction as is needed for the Moon, nor any equation of time as is needed for right ascension, where accuracy in timing to minutes or even seconds is vital.

Allan Chapman has conservatively estimated the accuracy of the original Mural Arc at Greenwich at 12" (Chapman 1982, p. 6), but as this historic instrument was lost on Flamsteed's death, this estimate was an inference, merely based on comparable instruments of the time (The disappearance of the Mural Arc followed Halley's filing of a lawsuit against Flamsteed's widow, claiming the equipment as his own on the grounds that he was the new Astronomer Royal. He lost the case, she lost the instruments).

Lunar readings on that Mural Arc could not aspire to quite such exactitude as the stellar positions. Taking a vertical reading as the lunar limb touched the central filament of the telescope's eyepiece was a less accurate affair. By the time the data had been tabulated and had certain astronomical adjustments applied, the errors would be greater. We found that, for a batch of 16 positions sent to Newton in Flamsteed's letter of February 7th, 1695 (reproduced in *Correspondence*, IV, p. 84) the mean error in longitude was -15 ±48 seconds of arc (Kollerstrom & Yallop 1995, p. 242). That was after Flamsteed had applied various corrections and converted from equatorial co-ordinates (declination, right ascension) to ecliptic (longitude, latitude). The issue of how lunar data was reduced into a form suitable for theoretical use will be reviewed later. Here we are merely concerned to make some preliminary comments about data accuracy.

II. Conditions of Composition

The conditions under which *TMM* was composed have a couple of rather strange, indeed startling, features. The original manuscript (kept at

Cambridge University Library, Add. 3966), was composed on February 27, 1700. Its date of composition comes from David Gregory, a reliable source because of the reverence with which he recorded matters Newtonian. His copy is in the library of the Royal Society, with the composition date marked. The "epochs" of *TMM* were noon on December 31st 1680 and 1700, the two limits over which various celestial positions were given. The positions for the latter date were predictions, as had not then been reached.

In the very month of *TMM*'s composition, Newton was confirmed as the Master of the Mint. On the third of February, a royal edict proclaimed:

> Know yee that wee for divers good causes…do give and grant unto Our trusty and Well beloved Subject Isaac Newton Esqr. the office of Master and Worker of all our Moneys both Gold and Silver within our Mint in our Tower of London and elsewhere in our Kingdom of England … And know yee that wee for the considerations aforesaid have given and granted, and by these presents do give and grant unto the said Isaac Newton all edifices, buildings, gardens, and other fees, allowances, profitts, privileges, franchises and immunities belonging to the aforesaid Office.

The stress of the great recoinage had passed away, and perhaps some new hope dawned that he could indeed resolve the problem. And yet, it cannot but strike us as rather extraordinary, that within weeks of acquiring such a responsible position, one of the most demanding jobs in the country, Newton should find time to ponder the niceties of lunar motion, and compose a brief but obscure opus on the subject. Not long after, Newton would have to ready himself to stand for the Trial of the Pyx, whereby the quality of the gold of the nation's currency was tested and to which the Master of the Mint was personally answerable for deficiencies. Not less than two thousand pounds was expected to be submitted by the Master of the Mint in advance as a security for the operation. His full attention was expected over the problems of bimetallism, whereby the differing values of gold and silver defined the relative weights of the currencies cast in them. In these years he still retained his position—and income—as a Fellow and Professor of Trinity College, Cambridge.

The twenty-year period specified by the epochs of *TMM* was a multiple of four, whereby the leap years would fit in and not disrupt the flow of computation, and was the smallest such multiple to embrace a Saros and nodal cycle, each of eighteen years. But, in addition, these two decades had a very personal significance for Newton. Without wishing to over-generalise, one notes that they framed the main period of Newton's creative life in relation to astronomy. In 1680 there arrived the great comet which the Trinity lecturer sat up observing, followed in 1682 by what was later recognised as Halley's comet. His composition of *TMM* in 1700 appears as the grand finale of that output. There is no real evidence of his further studies of the matter after this date (Baily 1837, p. 706). His vast ruminations on the cosmic process were framed by these two decades.

TMM's date of composition being controversial, there are three further occasions when textual evidence relevant to this will be treated: Ch. 4, III, discusses *TMM* mean motions from a separate document; Ch. 7, III, comments upon values for the apse equation values; Ch. 9, VIII, evaluates what has been alleged as an early draft of *TMM*.

III. Halley's Hope

TMM was published by David Gregory, formerly professor of Mathematics at Edinburgh University, who in 1702 became Savilian Professor of Astronomy at Oxford University. The title of Gregory's textbook, in which *TMM* was included, was (in English translation): *The Elements of Physical and Geometrical Astronomy.* The claim to have established a "physical" astronomy echoes that made earlier by Kepler, at the front of his *Astronomia Nova*. Introducing *TMM*, Gregory dismissed previous endeavours in this area for their lack of a physical basis:

> But as they made their Tables not from known Physical Causes
> and their Periods, but only by attending to Observations, it is
> no wonder if they did not rightly distinguish the Inequalities
> from one another." (Gregory 1715, p. 132)

This alludes to the large question of the extent to which *TMM* was based upon "physical causes," when nothing in its text indicated such. But, what did Gregory mean by claiming that his astronomy was also "geometrical"?

In a sense, "geometrical" merely signified, "perfect," alluding to the exact nature of geometrical proofs, as free from approximations. It echoed ancient Platonic notions about astronomy with which his readers would have been familiar. Later in the century it would become evident that, for matters involving time-dependent variables, geometry was not so suitable. Fluxional and differential methods were then just beginning to be adopted by mathematicans, and half a century later would become the new format for expressing these things.

A comment by Edmond Halley, made while discussing the *Principia*'s lunar section in its first edition, is worth quoting in this context:

And tho' by reason of the great Complication of the Problem, he has not been able to make it purely Geometrical, tis to be hoped, that in some further Essay, he may surmount the difficulty.

(This was from Halley's "True Theory of the Tides," *Phil. Trans*, 1697, 19, pp. 445-457, composed a decade earlier, for Halley's presentation of *PNPM* to James II: Armitage, p. 67; Cook p. 505). If it strikes us as curious today, it is because we view progress in this area as having taken place through the discarding of geometrical methods, and their replacement by algebraic functions. Our ability to believe that a historical figure was applying a theory of gravitation to deduce or obtain results, is likely to depend upon their having progressed in some degree beyond a merely kinematic or geometrical mode of reasoning. However, *TMM* in 1702 developed a geometrical mode of reasoning, just as Halley had hoped.

Gregory extolled the accuracy of *TMM* highly in an introductory paragraph, though it was a thing he had no means of assessing:

By this theory, what by all Astronomers was thought most difficult and almost impossible to be done, the Excellent Mr Newton hath now effected, viz. to determine the Moon's Place even in her quadratures, and all other Parts of her Orbit, besides the Syzygies, so accurately by calculation, that the Difference between that and her true Place in the Heavens shall scarce be two Minutes, and is usually so small, that it may well enough be reckoned only as a Defect in Observation.

(Cf. p. 5, above)

Gregory has here made the bold claim that the theory's predictions were only limited by the accuracy of the data on which it was based. It was what would nowadays be called a sales blurb. Gregory was a theoretical astronomer. Flamsteed referred scathingly to him as a "closet astronomer" because he did no practical work (Baily 1837, p. 204). We will see to what extent Flamsteed's opinion, at least over this specific issue, was justified.

On the other hand, Gregory's judgement was largely endorsed, years later, by the Astronomer Royal. After 1720 when Edmond Halley assumed that post, he had the opportunity to check *TMM* against accurate data. His opinion, which we have already in part quoted, shows the strongly politicised nature of the discussion, which seems to have continued ever since. Halley found that "for whole months together" *TMM* was:

rarely differing two minutes of Motion from the Observations themselves; nor is it unlikely but good part of that Difference

> may have ben the Fault of the Observer. And where the Errors were greater, it was in those parts of the lunar orb where Mr Flamsteed had very rarely given himself the Trouble of observing: viz, in the 3rd and 4th quarter of the Moon's Age, where sometimes these differences would amount to at least 5 minutes" (*Phil. Trans*, 37, p. 191).

My investigations have not confirmed either that errors of such magnitude were present in the lunar observations of Flamsteed, or that the data came mainly for the waxing half of the lunar orbit tending to omit the last two quarters. A later section will assess the question of data accuracy and reliability. It becomes a rather central issue, if both Whiston and Halley are claiming that the performance of *TMM* was limited primarily by the data on which it was based.

These days, the pendulum has swung in the opposite direction. Curtis Wilson boldly described Newton's great lunar endeavour as a "failure" (Wilson 1987, p. 76), and the reference cited for that claim was the Whiteside tercentenary essay. There is room for doubt as to whether Whiteside adopted quite so extreme a position. It is worth quoting the conclusion of D. T. Whiteside's tercentenary address, for this study has formed the starting point of modern discussions of the topic:

> It is, unfortunately, one of the most tenacious myths of Newtonian hagiography that this demi-god of our scientific past made his dynamical explanation of the moon's motion in all its irregularity the supreme proof of his monolithic principle of the universal inverse-square law of gravitation which governs all celestial and terrestrial movement, and this in a surpassingly rigorous geometrical manner which he made inimitably his own. "Who," to quote Whewell's eulogistic phrase of a century and a half ago, "has presented in his beautiful geometry, or deduced from his simple principles, any of the [lunar] inequalities which he left untouched?" The truth, as I have tried to sketch it here, is rather that his loosely approximate and but shadowily justified way of deriving those inequalities which he did deduce was a retrogressive step back to an earlier kinematic tradition which he had once hoped to transcend, and to a limited Horrocksian model which was not even his own invention (Whiteside 1976, p. 324).

More recently, Wilson concluded a fine study of the matter by saying, that the Newtonian lunar endeavour had come unstuck because:

Newton's effective adoption of Horrock's lunar theory, by interfering with ongoing insight into perturbations not actually embraced by that theory, proved ultimately an insurmountable obstacle to him (*GHA*, p. 267).

That is a novel interpretation of the failure, if indeed we should regard it as such. A great British discovery, which formed the backbone of the finest lunar theory available (that of Flamsteed, the Astronomer Royal), is blamed for having prevented a mathematican from having been more successful, by virtue of his adopting it. Perturbation theory is something one thinks of as developing in the middle of the eighteenth century, and in France. We are merely pointing out that a problem seems to exist, in deciding whether or not an enterprise was a success or a failure, and if the latter, on what that should be blamed.

In a sense such verdicts must be conjectural. Until we know how *TMM* functions, as an integral whole, it must remain so. Until then, we can only quote the radically opposed views of Halley and Flamsteed (for example), and perhaps side with one or the other. Here we aspire to reach beyond such an armchair approach, and resolve centuries-old controversies in a practical manner. *TMM* is like a machine, a watch, which once wound up and set in motion will generate positions for the luminaries. It will do this, provided only that we can follow its instructions. We here aim to set its antique wheels in motion, to see how they move one against the other, thereby to gain insight into what has long been an obscure and neglected area in the history of science.

IV. Moving the Goalposts

After the cognoscenti had been nodding their heads over these matters for three years, and rumours put about that such profound accuracy had now been achieved that Flamsteed need not bother any more in gathering lunar observations, for the job was done (Baily 1837, p. 176), Newton then submitted some "Corrections" applying to *TMM*, one of which shifted the position of its "mean moon" by ten arcminutes. He thereby displaced the values which *TMM* would generate by five times more than its supposed maximum error, as affirmed by Gregory. The concept of a "mean moon" and of this adjustment will be elucidated further in a later chapter, but suffice to say that it is the fundamental starting point for a lunar theory. It is hardly adequate to characterise such an alteration as an "correction." (Cohen 1975, p. 87. The "Corrections" appeared in *Miscellanea Curiosa* of 1705, published by the Royal Society.)

The year before these "corrections" appeared, Flamsteed described how the Royal Society's President paid him a visit at Greenwich. He was shown some early lunar positions of the 1670s and 80s, and their disagreements of up to ten minutes with *TMM*:

> I showed him also my new lunar numbers, fitted to his corrections; and how much they erred: at which he seemed surprised, and said "It could not be." But, when he found that the errors of the tables were in observations made in 1675, 1676, and 1677, he laid hold on the time, and confessed he had not looked so far back: whereas, if his deductions from the laws of gravitation were just, they would agree equally in all times. (Baily 1837, p. 217)

This was in a letter to Abraham Sharp. Sharp was a mathematics teacher with whom Halley corresponded over a formula he developed for the convergence of π, who built the Mural Arc in 1690 that was much admired, and who stuck the stars onto Flamsteed's maps of the constellations. He is thus a fairly significant witness to the course of events. Flamsteed may have been by this time (1704) rather embittered by certain aspects of Newton's behaviour, but that is no reason for dismissing or marginalising his opinions, as commonly happens in histories of these events. This altercation indicates the rather significant fact that Flamsteed reckoned that he was able to make *TMM* work, at the beginning of the eighteenth century. Had he composed new tables for it? We shall see later, that there is evidence for Flamsteed being the first person to do this.

V. A Modern Approach

We aspire to follow the path of a new generation of historians of astronomy, pioneered in America by Owen Gingerich. They have used computers to probe into the past to ascertain how accurate were the endeavours of any astronomer in history. Thereby they have given a greater emphasis on the practical side of astronomy in a historical context, which was much needed. Centuries-old discussions are now resolvable, and a precise new basis can be given to the history of astronomy.

Gingerich (at the Harvard-Smithsonian Center for Astrophysics) studied "error patterns" in ephemerides, by comparing their predictions against actual positions over the years. This showed the extent to which the theories of astronomers were succeeding in practice. If we knew of a lunar almanac which had used *TMM*, we could assess its accuracy simply by following this approach. Curtis Wilson has probed into the specific components of lunar theories of this period, comparing the diverging values of solar and lunar constants.

In Britain, the most accurate lunar and planetary programmes are those developed by the Royal Greenwich Observatory, at the Nautical Almanac Office. In the 1950s, the "Improved Lunar Ephemeris" was there developed, and revised in the 1970s, to obtain something near to one second of arc accuracy in historical time. It has around sixteen hundred terms for longitude, as compared with the historical theory we are examining which had seven. It is powerful enough to be able to determine the accuracy of the work of the founder of the R.G.O., the Reverend John Flamsteed. Three centuries after Flamsteed set up his great mural arc in 1691—characterised by Allan Chapman as "the finest and most exact astronomical instrument constructed to-date" (Chapman 1990, p. 57.)—computers can finally match its accuracy in checking its positional data.

Chapter 4

A Commentary on TMM

Our approach here is complementary to that of Bernard Cohen's excellent 1975 treatise. Cohen there discussed the circumstances of *TMM*'s production, its various editions in Latin and English, the comments made upon *TMM* by astronomers, and why science historians had largely ignored the subject. We on the other hand are concerned with the argument of this document. This has scarcely (we believe) hitherto been attempted. This Chapter deals with its first seven paragraphs. The remainder of the commentary appears in Chapter 9. Each paragraph of *TMM* is printed in boldface, then discussed. For modern values of some of the *TMM* constants, see Appendix I.

The Royal Observatory at Greenwich is to the West of the Meridian of Paris 2° 19'. Of Uraniburgh 12° 51' 30". And of Gedanum 18° 48'.

The *Paris meridian* is 2° 20' East. In those days there was no general agreement on the "Greenwich Meridian," so the distances between observatories in longitude were vital for comparing observations. For the once-glorious *Uraniborg*, then fallen into rack and ruin yet still important for astronomers as the site where "the prince of astronomers" as Flamsteed called him, Tycho Brahe, had once worked, the true longitude is 12° 27' East. *TMM*'s value for its distance in longitude erred by 24 minutes! This would have introduced an error of $1\frac{1}{2}$ minutes of time into any data that was being transcribed, from Uraniborg time to Greenwich time.

Gedanum referred to the observatory of Danzig (now Gdansk), where Hevelius worked. This would have grown into the most illustrious observatory in Europe, had it not tragically burnt to the ground in 1679. Flamsteed compared many of his observations with those of Hevelius, and was startled to find them agree to within a fraction of a minute in many cases, even though Hevelius used only his own eyesight unassisted by the new telescope plus micrometer-gauge. With such close agreement, the correct time-correction would have been a vital matter. Its correct longitude is 18° 24', so *TMM*'s position was again in excess, by 23 $\frac{1}{2}$ minutes of arc.

The mean Motions of the Sun and Moon, accounted from the Vernal Equinox at the Meridian of Greenwich, I make to be

> **as followeth. The last Day of December 1680 at Noon (Old Stile) the mean motion of the Sun was 9 sign 20° 34′ 46″. Of the Sun's Apogaeum was 3 sign 7° 23′ 30″.**

Motions signified "positions" at specified epoch times, measured in ecliptic longitude, and cited for noon as the time of day for which an ephemeris had to define positions. The zodiac begins from 0° Aries at the Vernal Point, so *9 sign* meant the nine zodiac signs on from that position on the ecliptic, viz the sign of Capricorn. In the next section we consider how accurate was the Newtonian value for the Sun's mean position. Gingerich found that solar errors in ephemerides of this period were not more than several minutes of arc (Gingerich 1983, p. xix). Solar positions were straightforward to calculate, depending merely upon the eccentricity value used for the Earth's orbit.

The *Sun's apogaeum* referred to the Sun's position in longitude when the Earth was furthest away, at midsummer. It utilised the old geocentric terminology, whereby the Sun was seen as circling around the Earth. The more or less fixed position of the aphelion is here specified to an accuracy of four minutes.

> **The mean Motion of the Moon at that time was 6 sign 1° 35′ 45″. And of her Apogee 8 sign 4° 28′ 5″. Of the Ascending Node of the Moon's Orbit 5 sign 24° 14′ 35″.**

We have seen how the *mean node motions* of the period had an accuracy of two minutes (Ch. 1, V). As regards the accuracy which Newton hoped his *mean moon position* to have, we may quote from a letter of his to Flamsteed of January 15, 1695—the period of early optimism:

> In trying to compute the mean motion of the moon from the *tempus apparens* in some of your observations, I find that the mean motion, gathered by my computations, differs sometimes from that in your synopses 5″ or 6″, or above. Which makes me suspect that, in determining the *tempus apparens*, your servant followed some tables which are not sufficiently exact.

Years later, some "corrections" were specified for *TMM*, (Cohen 1975 p. 87) one of which was a ten minute displacement of the mean moon, from 1° 35′ 45″, to 1° 45′ 45″. The new value was used in Whiston's reprinting of *TMM* in 1707. It does better relate to the final lunar position given for 1700, in terms of the mean tropical lunar period which links them together.

Five seconds or ten minutes? Years later Flamsteed concluded that 1′ 30″ should be added onto the Newtonian value for mean moon (August 31, 1714, Baily 1837, p. 698). Chapter Five will ascertain to what extent he was correct on this matter.

Lunar apogee here refers to one end of a notional apse line, which had uniform motion in longitude, not the actual position of apogee (Ch. 2, III). This apse position is cited as being 244° 28′ 5″ for the epoch. A modern estimate of its true position was about three minutes more than this (Ch. 5, V), making it quite an accurate value for the period. The apse line position was the foundation for lunar theories.

And on the last Day of December 1700 at Noon, the mean Motion of the Sun was 9 sign 20° 43′ 50″. Of the Sun's Apogee 3 sign 7° 44′ 30″. The mean Motion of the Moon was 10 sign 15° 19′ 50″. Of the Moon's Apogee 11 sign 8° 18′ 20″. And of her ascending Node 4 sign 27° 24′ 20″. For in 20 Julian Years or 7305 Days, the Sun's Motion is 20 revolut. 0 sign 0° 9′ 4″. And the Motion of the Sun's apogee 21′ 0″.

These positions are mechanically linked to the previous ones of 20 Julian years earlier, through their mean periods. The *Sun's position in longitude* has moved on by 9 arcminutes, because the Julian year of the calendar then in use of 365.25 days is not quite the same as the Tropical year, whereby it generates that displacement over 20 years. Britain was at this time refusing to abandon the old Julian calendar, as most of Europe had a century earlier, for religious reasons. Instead of zodiac signs, *TMM* counted the number of signs from the vernal equinox, for example 20° Capricorn would be expressed as 9s 20°.

The sun and moon are mathematical abstractions moving uniformly, as will normally differ from the true positions. The goal of a lunar theory was to bridge that gap, which evidently could be as much as six degrees, and to do so within arcminutes. It is of interest to compare these mean values with the actual positions at the time, i.e., the modern-calculated values, to give an idea of what was involved:

	Mean position		True position		Difference
Epoch date 1680					
Sun	20° 34′ 46″	Cap.	21° 1′	Cap.	26′
Moon	1° 35′ 45″	Lib.	8° 3′	Lib.	6° 27′
Node	24° 14′ 35″	Vir.	24° 17′	Vir.	2′

	Mean position	*True position*	*Difference*
Epoch date 1700			
Sun	20° 43′ 50″ Cap.	21° 10′ Cap.	26′
Moon	15° 19′ 50″ Aqu.	16° 59′ Aqu.	1° 39′
Node	27° 24′ 20″ Leo	27° 27′ Leo	1° 3′

Two months after *TMM*'s composition, in April of 1700, we find the Master of the Mint again musing upon *Elementa motuum Solis et Lunae ab Aequinoctio verno*. He gave mean positions for January 1 1701, Old Style (*Corr.* IV, p. 328), a mere day later than *TMM*'s epoch, as conveniently enables us to inspect his differences:

Mean positions for Jan.1st, 1701	*Diurnal Motion*
Sun 21° 42′ 38″ Cap.	58′ 48″
Moon 28° 30′ 12″ Aqu.	13° 10′ 22″

The difference-values fall slightly short of *TMM*-based diurnal mean motions: 59′ 08″ and 13° 10′ 22″ respectively, in deficit by 20 and 12 arcseconds, suggesting that Newton could have been mulling over slightly different mean-motion values. But this data surely supports the date given by Gregory of early 1700 for *TMM*'s composition. Normally, one would not require confirmation of so reliable a source; however, it was as we have seen a rather extraordinary period in life to choose an attempt to fathom this highly inscrutable issue, and we may be grateful for supporting evidence over its date of composition.

The *mean solar motion* given in *TMM* is equivalent to one revolution per tropical year of 365.242 days, just as its *mean lunar motion* is equivalent to one revolution per tropical month. This should not be confused with its use of a Julian calendar, as took exactly 365.25 days per solar year. This calendar had one leap year in four intercalated with no other adjustments.

The figure given for the *motion of the sun's* [i.e., earth's] *apse line* is of interest, as the motion, or "quiescence" as *PNPM* put it, of the planetary apses was controversial. Vincent Wing in his *Urania Practica* (1649) gave it a motion of 1′ 01″ per annum, somewhat less than *TMM*'s value of 21′ in 20 years. Streete in his *Astronomia Carolina* said the planetary apses were immobile with respect to the stars, meaning that their motion against the zodiac was identical with the precession value. Flamsteed in the preface to his *Historia Coelestis* suggested 1′ 3″ as the annual motion of the Earth's apse line with respect to the Equinox (Chapman 1982, p. 147), which is identical to the *TMM* value. Sidereally, the apse moves 11".8 per annum, and so adding the Vernal Point's motion of 50".2 gives its tropical motion of 1'2" per annum.

William Whiston, in his astronomical lectures published in 1707, expressed surprise at *TMM*'s putting the Earth's apse in motion. The notion that the planetary apses moved had been "exploded" out of astronomy, he remarked (Cohen 1975, p. 149), and so why were they here brought back again?

**The Motion of the Moon in the same Time is 247 Rev. 4 sign 13°
34' 5". And the Motion of the Lunar Apogee is 2 Revol 3 sign 3°
50' 15". And the Motion of her Node 1 revol. 0 sign 26° 50' 16".**

The length of the tropical lunar month is here indicated. A figure of 247 revolutions was accidentally given, in this paragraph and the next, which was corrected in 1705 (Cohen 1975, p. 87) to read 267. An exact value then emerges. Dividing 20 Julian years by the number of revolutions here specified gives a mean period of the tropical lunar month within one-fifth of a second (Appendix I). It must have been the most accurately known physical constant at that period. Though *PNPM* only cited the sidereal lunar month within the nearest minute, *TMM* used its mean duration to a second.

**All which Motions are accounted from the Vernal Equinox:
Wherefore if from them there be subtracted the Recession of
Motion of the Equinoctial Point in antecedentia during that
space, which is 16' 0", there will remain the Motions in reference to the Fixt Stars in 20 Julian Years; viz. the Sun's 19
revol. 11 sign 29° 52' 24". Of his apogee 4' 20". And the Moon's
247 revol 4 sign 13° 17' 25". Of her Apogee 2 revol 3 sign. 3° 33' 35".
And of the Node of the Moon 1 revol 0 sign 27° 6' 55".**

A procedure for converting from tropical to sidereal space is here given. The reference framework becomes that of the fixed stars, no longer a moving zodiac system anchored to the Vernal Point. This move has theological implications, because sidereal space was the sensorium of the Deity for Newton, and the ascent into that reference framework, where the centre of mass of the solar system was immovable, away from the merely human perspective on things, was for him a religious exercise, or so he declared at the start of the *PNPM* (pp. 52–3; cf. pp. 761–4). For now we merely note that the 30° signs here referred to are sidereal, that is pertaining to that zodiac system invented by the Chaldeans and defined by fixed stars. *TMM*'s instructions pertain to two different zodiac systems.

Against this immobile sidereal space, the monthly orbit of the Moon, used with such remarkable success in *PNPM* to show that gravity reached as far as the lunar sphere, is here ascertained to a few parts in ten million.

Astronomers required the ability to make such a conversion, from tropical to sidereal longitude, as the positional data could well be given with respect to a fixed star. There was however no generally accepted sidereal reference framework. An inaccurate value for precession is here given, of 16′ 0″ in 20 Julian years, corrected in 1705 to 16′ 40″, which is one degree in 72 years.

According to this Computation the Tropical Year is 365 days 5 hours 48′ 57″. And the sydereall Year is 365 days, 6 hours, 9′ 14″.

This concludes the dualistic system introduced in the previous paragraph, whereby two distinct reference frameworks are introduced, enabling the reader to switch over to the sidereal, and back to tropical. The values are more accurate than those cited in Streete's *Astronomia Carolina* of 1661 (Streete 1661, p. 45). The period of the sidereal year is given correct to five seconds, and that of the tropical year to ten.

Chapter 5

Mean Motions of the Sun and Moon

To construct a lunar ephemeris, five "mean motions" were required: of the Sun, Moon, apogee, aphelion and lunar node. The aphelion was almost stationary, only moving a degree or so per century, so there were to all intents and purposes four positions which had to be located on the zodiac for any given time, as one's starting-point. Their accuracy could easily limit the accuracy that an ephemeris achieved: if a mean moon was out by several arcminutes, its predicted positions would err by that amount on average. How accurate were the mean motions of *TMM*, and were they better or worse than others of the period? We here answer these questions, using the modern equations.

A "mean Moon" traverses its orbit with uniform angular velocity. Thereby it meets the "true Moon" twice-monthly, at or near to the mean apogee and perigee. A week after perigee the true Moon will lead the mean moon by about six degrees, while a week after apogee it will lag behind by a comparable distance. Such mean angular velocities were then called "mean motions," and had their ecliptic longitudes specified at twenty-year interval epoch dates. Somewhat confusingly for us, this term "mean motions" would often be applied to these epoch positions.

Such "mean motions," i.e., defined mean positions at twenty-year intervals, provide a simple first step in ascertaining whether tables used by ephemeris-makers were "Newtonian." Numerous tables in the first half of the eighteenth century claimed to be so, and Sir William Whewell averred in 1837 that *TMM* was:

> for a long period the basis of new Tables of the Moon, which were published by various persons; as by De L'Isle in 1715 or 1716, Grammatici at Ingolstadt in 1726, Wright in 1732, Angelo Capelli at Venice in 1733, Dunthorne at Cambridge in 1739. (Whewell 1837, II p. 209).

The issue was discussed by Baily (1835, pp. 701–705), and more recently by Craig Waff (in Cohen 1975, p. 79) and Curtis Wilson (*GHA*, pp. 267–8). Newton gave *four* successive versions of his mean motions, thrice modifying that given in *TMM* of 1702, so there were options as to which of these a "Newtonian" ephemeris adopted.

Our concern is not with yearly-produced ephemerides as gave daily tables for the positions of the heavenly bodies (for which see Kelly 1991), but rather with the mean motion tables as were required to construct them. The French *Connoissance des Temps* which ran from 1678 onwards was a fine example of the former, probably used by Halley in his South Atlantic voyages.

I. The Modern Equations

Mean motion implies a concept of uniform angular velocity without any periodically-varying terms. It is an average rate of movement in ecliptic longitude, computed by modern convention using Julian time. The time-value is measured in Julian centuries backwards from a given epoch, which is nowadays AD 2000. Its computation requires three terms: a constant representing the starting point, plus a T and a T^2 term. We ignore higher-power terms, their effect being less than an arcsecond.

There are some long-period terms that have an amplitude of around ten arcseconds. There is a case for not including such periodic terms, if one is aiming to calibrate historical conceptions of mean motion. They are not used in the mean motions of the modern theories of Meeus and Chapront-Touzé, though they were used in the older theories of Brown and Newcomb. It would be an option to include them here, and would displace the error-values estimated in this chapter by some arcseconds.

Slight disagreement continues to exist over modern equations for mean motion, with a cross-channel divergence of opinion, but only in arcseconds. It seems likely that the French tables are to be preferred: the new issue of the *Explanatory Supplement for the Astronomical Ephemeris* (1992) has mean motions slightly diverging from those of Jean Meeus, and they have not been revised since the earlier edition. The divergence is somewhat *larger* than that between the historical tables and the Meeus values, especially for the Sun (See Appendix II).

The computations here performed would have been been considerably less accurate if attempted earlier, prior to their time of composition in 1992. The new copy of Dr Meeus' book, *Astronomical Algorithms* (1991), incorporates the improved parameters of Michel Chapront-Touzé and Jean Chapront (1988), resulting from high-precision dynamical studies of earth-rotation, and has notably improved the secular variation terms for the five variables that concern us. Thanks to the work of the Chapront-Touzés, one is for the first time able to go back into past history, with terms probably accurate enough to assess the accuracy of astronomical mean tables for centuries gone by.

By way of indicating the improvement that has come about, Appendix II shows the divergences in mean lunar motion estimates from several sources for integer Julian centuries. Meeus' 1986 textbook on positional astronomy used older formulae based upon E. Brown's lunar theory, and was found to be considerably less accurate than his new algorithms (1991).

In the case of the Sun's motion, an accuracy of an arc second or two is required*. The mean motion formulae give Dynamical Time (formerly Ephemeris Time), which for historical investigation must be translated into Universal Time. The difference is due to Earth's deceleration over the centuries, caused by tidal friction from the Moon's pull. The irregularly varying difference between the two kinds of time is ΔT. ΔT is added on to the time function before the computation. For historical studies the equation may be expressed as:

$$\text{Ephemeris time} = \text{GMT} + \Delta T.$$

Modern astronomers retain Julian time for studying the past, and to every date assign a "Julian day number." They measure Julian time from noon on December 31, 2000, and not 1900 as was earlier done. Thus, for an epoch value in 1700 New Style one enters $T = -3.0$ into the time-equations. To convert a Julian Date into this form, where time is measured in centuries, the equation is:

$$T = (\text{JD} - 2451545)/36525 \text{ J.centuries,}$$

where 2451545 is Julian date of AD 2000 epoch and 36525 the days in a Julian century.

---

* For lunar positions the conversion from UT (Universal time) into TDT "Terrestrial dynamical time" is relevant (TDT emerged in 1984 to replace Ephemeris time, ET), due to its larger daily motion. Ephemeris time had been a measure of what was supposedly uniform time, derived "from the uniform motions of the planets," while the somewhat misleadingly called Universal Time remains "defined by the rotational motion of the Earth" (Meeus date, p. ???), having replaced GMT in this context in the 1930s. The latter is subject to variations which are "unexpected and unpredictable" (McNally, 1974, p. 88).

For the development of TDT, Terrestrial Dynamical Time, out of Ephemeris Time, giving $\Delta T = \text{TDT} - \text{UT}$, see *The Astronomical Almanac* 1993 pp. B4–B7. Until recently its equation was expressed as $\Delta T = \text{ET} - \text{UT}$. Tables in the Explanatory Supplement to the Astronomical Ephemeris (1992, pp. K8–K9) give this variable ΔT from 1620 onwards, as being sometimes as little as five or ten seconds of time, more or less, but as over a minute for the early seventeenth century.

These mean-motion computations can involve multiplying two ten-figure numbers together! For example, the modern convention gives Newton's epoch value of noon, December 31, 1680 Old Style a value of

-3.189650924 Julian centuries.

If only nine figures are used for this function, it will give erroneous seconds of arc positions. The term by which it is multiplied, derived from the tropical-month period, is also specified to ten figures. Astronomers used to split up these long numbers into two parts for this sum; however, nowadays a computer has no problem in multiplying them together. In addition, a "modulus" function is required to give the answer as a number between zero and 360°; for example, "mod(730,360)" gives 10°, as the remainder after division.

II. Newtonian Values

The historical tables cite mean motions at twenty-year intervals. When such a position was cited for a date, say for 1701, it referred to the noon on the last day on the previous year, as remains the practice to this day. As Flamsteed expressed the matter,

> The Radices of the mean Motions are fitted to the Meridian of
> London, and the Noon preceding the first of January.
> (Flamsteed 1681, p. 33)

For example, Whiston's *Lectures on Astronomy* (Whiston 1707) had tables with epoch values for the year 1681. These gave mean sun, moon, apse and node positions for December 31, 1680 identical with those specified in *TMM*, as being indeed the first textbook to use them. *TMM* cites its positions as for noon and leaves the reader to assume that mean time rather than apparent is intended. Comparing these epoch values from *DOS* and *TMM* (as corrected in 1705) with the modern Meeus values, as derived from Chapront-Touzé (1988 p. 346):

1681 Mean Epoch Positions (London):

	Lunar	Apogee	Node
DOS (1681)	181° 43′ 58″	244° 11′ 51″	174° 14′ 33″
TMM (1705)	45′ 45″	28′ 05″	14′ 35″
actual (for Greenwich)	45′ 46″	30′ 53″	17′ 6″
TMM errors:	-01″	-02′ 48″	-02′ 31″

	Solar	*Aphelion*
DOS (1681)	290° 34' 48"	186° 51' 40"
TMM (1705)	34' 46"	23' 30"
actual (for Greenwich)	34' 51"	27' 24"
TMM errors:	-05"	-03' 54"

We have here taken the "Meridian of London" as five arcminutes due West of Greenwich. In the seventeenth century, London rather than Greenwich was the prime meridian for British tables. The adjustment is one-third of a minute in time.

A substantial improvement is here apparent over the two decades from *DOS* to *TMM*. Newton's 1680 mean happened to be within an arcsecond of the correct value, which was possibly fortituous. William Whiston took the first version of the mean motions, as given in *TMM* of 1702, not using the modified means that appeared in the *PNPM*'s Second Edition of 1713 (see below). As a Fellow of the Royal Society he was aware of the corrections that emerged with a 1705 edition of *TMM* (*Miscellanea Curiosa* 1705), but others were not so fortunate. Certain lunar-ephemeris constructors failed as we shall see to note this edition of *Miscellanea Curiosa*, thereby acquiring a ten minute arc error in lunar motion.

The epoch mean motions, for noon GMT on December 31, 1700, were also modified in the *Principia*'s second and third edition (III 35 Scholium, *PNPM* pp. 663–4). Their values compared with *TMM* are as follows:

1701 means:	*Lunar*	*Solar*	*Apogee*	*Node*	*Perihelion*
TMM (1702)	15°19'50"	20°43'50"	8°18'20"	27°24'20"	7°44'30"
PNPM (1713)	15°20'00"	20°43'50"	8°18'20"	27°24'20"	7°44'30"
PNPM (1726)	15°21'00"	20°43'40"	8°20'00"	27°24'20"	7°44'30"
"true" means	15°20'23"	20°44'04"	8°19'49"	27°27'19"	7°48'04"

Each of them lags behind the modern estimate of correct value, at least prior to 1726. The node and perihelion remain unaltered, being out by two and four minutes respectively, and the apogee value is improved in the Third Edition, whereas the Sun's error has almost doubled. The Newtonian lunar mean motions contained errors in the region of half an arcminute. The Third Edition of the *Principia* concluded this section remarking, "the mean motion of the moon and of its apogee are not yet obtained with sufficient accuracy." From comparing these 1700 epoch means, we can observe that the 1705 value was adopted by Whiston in his *Praelectiones* of 1707,

Dunthorne's *Practical Astronomy* of 1739, and Wright's *New and Correct Tables* of 1732, while the the the improved 1713 value was adopted by Leadbetter's *Complete System of Astronomy* of 1742 and Halley's *Astronomical Tables* of 1752. No one used Newton's final 1726 value, as would then have been preferable.

To construct an ephemeris one needed an estimate of the lunar revolutions performed (in zodiac longitude) in twenty Julian years, supposedly accurate to arcseconds, for one's tables of mean motion. This interval was crucial, as any error would be cumulative. Values from some major sources are as follows:

Textbook	*Mean lunar motion per 20 years*		*Error*
Wing 1669:	267 rev., 133° plus:	33′ 44″	−57″
DOS (1681):		33′ 46″	−55″
TMM & Halley 1749:		34′ 5″	−36″
PNPM (1726):		35′ 15″	+35″
Cassini 1740:		33′ 58″	−42″

Deviations from the then-correct value are given to the right, by comparing with the Meeus-Chapront-Touzé equations (1988 p. 346), which only altered by an arcsecond or so over this period. *TMM's* value was accurate to one part in 10^7 - an error sufficient to give, by the 1730s and '40s, a two arc minute error. Two minutes of arc was the accuracy required to claim the longitude prize, enabling terrestrial longitude to be determined within one degree, so this was a large error.

It was generally only in the 1730s and 1740s that lunar tables came to be based upon *TMM*, i.e. it took three decades for *TMM* to be put into practice. To quote the Victorian astronomer Baily,

> It appears that a period of more than 30 years elapsed before Gregory's Newtonian rules [Baily's somewhat perjorative term for TMM] were thrown into the form of tables for public use. (Baily 1837, p. 702)

Was there a school of lunar-position astronomers, in the early decades of the eighteenth century, who based their work upon *TMM*, as Dr Craig Waff has affirmed? If so, a simple criterion will detect them. Those who may be called "the Newtonians" took their twenty-year epoch values for mean lunar motion as identical with that of *TMM*. This simple criterion yields the following rather impressive list of published tables:

Whiston	1707		Dunthorne	1739	
DeLisle	1716	Paris (unpublished)	Brent	1741	
Grammatici	1726	Ingolstadt	Leadbetter	1742	
Wright	1732		LeMonnier	1746	Paris
Capello	1737	Venice	Halley	1749	

For the textbooks here consulted, see Appendix II. In addition to these, Peter Horrebov's *Nova Theoria Lunae* published in 1718 in Uppsala also described itself as Newtonian. Citing its mean motions as from Copenhagen, it gave one set for the epoch 1700 "which agree with Newton's," and it contained no tables (Horrebov 1718, pp 34–46). Also, Nicholas DeLisle's unpublished lunar tables in the archives of the Paris Observatory satisfy the above criterion (Delisle, *Tables du Soleil & de la Lune*).

These astronomers concurred to within a single second of arc per twenty years, in the above-defined parameter. The values they took for mean positions varied somewhat, but in their twenty-year intervals they were identical. As mentioned, Halley and Leadbetter added ten arcseconds to their mean position tables, through adopting *PNPM*'s 1713 values. This list hints at a rather wide impact made by *TMM*, as Dr Waff has claimed.

III. A Century of Mean Motions

The mean lunar motion specified by *TMM* was somewhat more accurate than those hitherto published. The trouble was, that it was mainly adopted several decades later, as we have seen, by which time it had accumulated an error of two minutes and so was no better than others. Cassini's tables were superior at the time when *TMM* began to be used. Figures 1–5 illustrate the situation.

The graphs in this chapter use three sets of 20-year epoch values selected from each table, centred around their time of publication. The differences between these values and the Meeus/Chapront-Touzé value of the mean position were plotted, with corrections added for local time where necessary. The errors were thus {historical values − modern values}. The procedure for deriving the modern estimates of the historic positions is given in Appendix II.

Each diagram contains the data from just six tables, although some more are included in Table 5.1. The French tables were in New Style, and so after February of 1700 they were eleven days ahead of the British. The tables of Le Monnier and Cassini had epoch values for 31 Dec 1699 rather than

31 Dec 1700 as for the English ephemerides. I found that there were 6929 days between the Newtonian epoch date of 31 Dec 1680, (i.e., 11 Jan 1681 NS) and this French epoch date. Timezone adjustments were made for Paris (Cassini and Le Monnier), Venice (Capello), Bologna (Riccioli), and Belgium (Van Lansberge), see Appendix III.

Errors in Mean Motion, 1650-1750

		ΔMoon	ΔSun	ΔApogee	ΔNode	ΔAphelion
Morinus	(1650)	0'14"	−6"	27'	+6'	−21'
Wing	(1651)	+3'12"	−8"	−6'	−3'	−22'
Shakerley	(1653)	+1'22"	−21"	−5'	−	−34'
J. Newton	(1657)	+1'22"	−42"	−9'	−	−33'
Street	(1661)	−	−	16'	+8'	−
Riccioli	(1665)	+10"	−2'8"	−	−15'	+1°34'
Flamsteed	(1681)	−1'59"	−3"	−19'	−3'	−36'
La Hire	(1687)	+2'57"	−40"	−45'	−3'	−1'
Greenwood	(1689)	−	−	−16'	−8'	+53'
Whiston/*TMM*	(1710)	−33"	−14"	−2'	−3'	−8'
Grammatici	(1726)	−1'12"	−23"	−0'.1	−4'	−3'
Halley	(1749)	−58"	−38"	+2'	−4'	−10'
Capello	(1738)	+10"	−41"	+1'	−4'	−3'
Cassini	(1740)	−1'57"	−20"	+5'	−1'	−11'
Leadbetter	(1742)	−2' 0"	−32"	−3'	−5'	−4'
Le Monnier	(1746)	−33"	−22"	+3'	−4'	−3'

Table 5.1: publication dates are cited, and mean motion errors computed (by subtracting the Meeus/Chapront Touzé values) for the twenty-year epoch nearest to that date (Appendix II). Errors are cited in arcminutes for the node, apse and aphelion positions, and to arcseconds for the luminaries. For the British texts prior to 1700, London mean time was used. Missing values indicate that they were not as such given, probably because the tables gave anomaly values ($M − A$, $S − H$). (For the meaning of these letters, see p. 99 below.)

The textbooks were located mainly from Gingerich and Welther (1983), Curtis Wilson (*GHA*), and a collection sent by Dr Craig Waff, (Jet Propulsion Laboratory, Pasadena). Dr Waff had intended to review these tables (Cohen

1975, pp. 77, 79) but instead his doctorate dealt with apse line motion (Waff 1975). For textbooks consulted see Appendix I.

Some tables did not give values for mean solar and lunar motions, but rather gave mean anomaly values (see next chapter) plus the apogee, so that anomaly had to be added onto the mean apse position to obtain the mean position. This generated positional errors, as the apse and aphelion mean motions were an order of magnitude less well determined than those of the luminaries. Our table has omitted such.

The table cites a roughly chronological order of publication, modified somewhat by the range of usable mean epoch values. Error values for the 20-year epoch were usually nearest the date of publication, e.g., Morinus' "Tabulae Rudolfinae" were published in 1650, and I have here taken his 1660 epoch, though I have only been able to locate these as republished in the second edition of Streete's opus of 1705 (cf. Streete 1661). The mean tables of Shakerley and John Newton only went up to 1660, so had to be centred on 1640 (Shakerley 1653; John Newton 1657, p. 29). The only solar-lunar tables published in the twenty years following *TMM* seem to have been those by Whiston (1707).

The Table has its first two columns for the luminaries in arcseconds and the other three in arcminutes. It relates ancient and modern definitions of mean motion. Averaging these errors irrespective of sign, and comparing these means for the eighteenth and seventeenth centuries, shows a general drift of improvement:

	Sun	Moon	Node	Apogee	Aphelion
Mean error (arcminutes):	0.5	1.3	5	16	22
% improvement C.18/C.17:	20%	35%	45%	85%	85%

Gaps in the Table indicate a problem with locating or interpreting the relevant data. Lansberg's tables (1632 pp. 10–50) did not seem to contain the twenty-year epoch values. The "Newtonians" Wright (1732 p. 27), Dunthorne (1739) and Brent (1741) have not been included, having identical tables to those of Whiston. Halley compiled his tables around 1720, as were only published posthumously, so his mean motions are cited for 1720. That has made his errors appear different from Leadbetter's, though their tables were identical.

IV. The Missing Flamsteed Tables

Shortly after *TMM* appeared in 1702, the Astronomer Royal Flamsteed expressed his disapproval, and set about constructing tables of his own, claiming that these would give better positions (Baily 1837, p. 211). The terms of his employment drawn up by Charles II mandated him to this task. His new tables were "40 quarto pages and upwards" he told Abraham Sharp. His letters to Sharp described them, explaining why they occupied "so many pages," adding that Sharp should feel free to tell the world that they had been drawn up, for "I desire to have them published as soon as may be" (Baily 1837, p. 212). Nothing further was heard of these documents— until, decades later, a Frenchman Lemonnier claimed to have them.

No trace of Flamsteed's decades of work on lunar theory appeared in the three bulky volumes of his *Historia Coelestis Britannica* which emerged posthumously in 1725. Pierre Le Monnier's *Institutions Astronomiques* of 1746 included the claim (p. 155) that his tables were both new and based upon those of the English astronomer Flamsteed. A letter by Flamsteed's co-worker Hodgson confirms this, discussed by Baily somewhat inconclusively (Baily 1837, p. 704). More recently, Curtis Wilson (*GHA*, p. 201) averred that Halley had given Flamsteed's tables prepared in 1702 to Le Monnier. If so, we can only wonder how Halley came to possess these vital documents, not published in his "pirate" edition of Flamsteed's *Historia* of 1712 (Halley 1712). They would be most significant for evaluating Flamsteed's achievment as the Astronomer Royal. Table 5.1 shows that Le Monnier's mean positions were of a high standard.

For the Flamsteed tables to have migrated across the Channel in this manner, from Greenwich to Paris, three steps of transformation would have been required: a nine-minute difference for longitude, an eleven-day calendar change to New Style, and finally a year's difference for the mean motions owing to a difference in a convention in presentation. We would have to assume Le Monnier performed these adjustments, though the Paris Observatory archives have no record of these manuscripts.

V. Mean Motion Graphs

A graphical approach facilitates insight into who copied from whom, necessary to evaluate the extent of *TMM*'s influence in this field. In the following graphs *TMM* is represented by William Whiston's opus of 1707, since his tables were the first to embody its mean motions. Each line spans a forty-year period, over three twenty-year epochs, the middle one being that whose error was given in Table 5.1. Source data is given in Appendix III. The plotting of three points in this manner also served as a check upon my arithmetic procedure.

The solar mean motions show a common downward slope, 18 arcseconds per forty years in the case of Whiston. As Chapter Four noted (Section VIII), Newton's period for the tropical year was in excess by ten seconds, in which time the Sun would move ahead by approximately 10/24 arcseconds, or seventeen arcseconds in forty years. The downward slope thus represents that tropical year error. LeMonnier's means look somewhat as if he

Figures 5.1 & 5.2: These graphs show "errors" in tables of mean motions in tables published 1650–1670, for mean Sun and Moon positions. For three sets of twenty-year epoch values, the values derived from the Meeus/Chapront-Touze equations (1988 p. 346) were subtracted from the published epoch values.

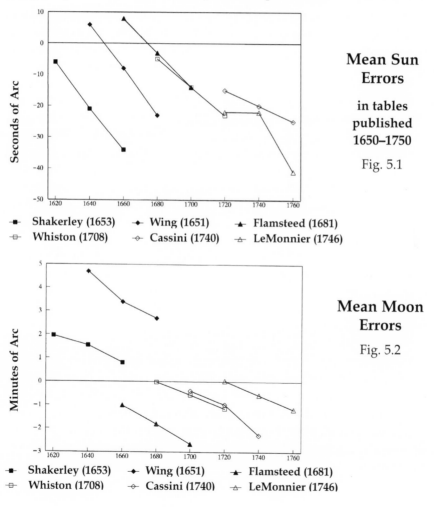

Mean Sun Errors

in tables published 1650–1750

Fig. 5.1

- Shakerley (1653)   Wing (1651)   Flamsteed (1681)
- Whiston (1708)   Cassini (1740)   LeMonnier (1746)

Mean Moon Errors

Fig. 5.2

- Shakerley (1653)   Wing (1651)   Flamsteed (1681)
- Whiston (1708)   Cassini (1740)   LeMonnier (1746)

Figures 5.3, 4 and 5: as before, for apse, lunar node and aphelion mean motions.

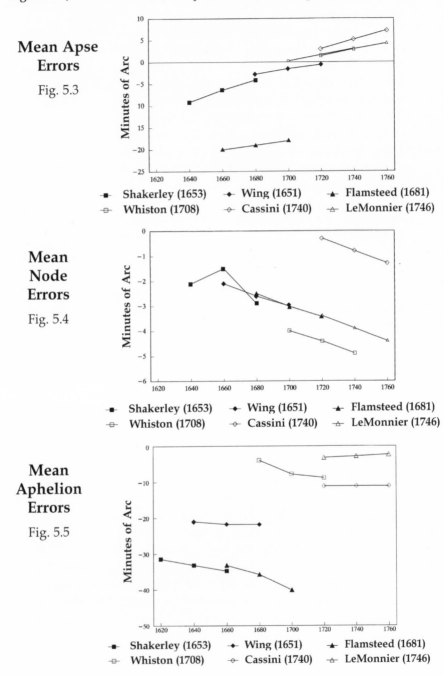

Mean Apse Errors

Fig. 5.3

Mean Node Errors

Fig. 5.4

Mean Aphelion Errors

Fig. 5.5

were using the *TMM* values, or trying to, while Cassini's can be seen as an improvement.

Chapter Four saw how *TMM* cited the tropical lunar month correct to 0.2 seconds, as the most accurately known physical constant at that period. Here, we see the cumulative effect of that 0.2 of a second, whereby over four decades it generated an error of about one arcminute. (Though historical mean motions were linear, the graphs are not necessarily straight lines, due to non-linear terms in the modern equations.)

The twenty-minute apse error in Flamsteed's 1681 publication is remarkable. In the year 1673 he caused Jeremiah Horrocks's theory to be published, whose equation of apse motion was its most remarkable feature. Flamsteed wrote, in 1675, of how he discovered the truth of the Horroxian theory:

> when I had found by many curious and careful measurements
> of the Moon's diameter, that the heavens would never admit
> those [earlier] Hypotheses. (*Phil. Trans.* 10, pp. 368-372)

referring to the pre-Horrocksian theories. His skill with a micrometer screwgauge would have to some degree helped him in locating the apse position. However, that same year Flamsteed published a criticism of an opus Thomas Streete had just published, for the way it claimed to be based upon Horrocks, in which he said:

> Mean time, when he hath done what he can (with his apse
> equation), it will not shew the true place to half a Degree.
> (Flamsteed 1674 p. 220.)

His own values for the mean apse were hardly better.

VI. Summary

Whiston's mean values, representing those of *TMM*, are as good as any in Table 5.1. Over the century a large improvement in the mean apogee and aphelion values appears, plus a smaller one for the lunar nodes. The graph shows Jacques Cassini's lunar means as more accurate than the Newtonian ones, reminding one of Owen Gingerich's account of how Paris in this period became the world centre for ephemeris contruction (Gingerich and Welther, 1983, p. xi), though Cassini was a generation later than Newton. The graphs emphasise a major feature of the Table, whereby historic mean values mainly lag behind what are nowadays regarded as their correct values. All five of *TMM*'s means, for both of its epochs, fall behind the modern values; except that, as the nodes are moving in the opposite direction, their historic values could be regarded as being in advance!

The mean motions here examined are one method of comparing the "flow of expertise" in constructing ephemeris tables. Another approach would be to compare the terms used in the various equations. For example, Le Monnier observed concerning the Equation of Apogee:

> La plus grande Equation du lieu de l'Apogée avoit éte établie autrefois par Flamsteed de 11° 47′ 22″. Mais M. Newton l'a augmenté & s'est assuré qu'elle devenoit plus conformé aux Observations lorsqu'on la suppose de 12° 18′. (Le Monnier 1746, p. 191)

This shows a remarkable degree of interest in the English method of computing the secondary apse motion, four decades after the theory was developed. The French were then unfamiliar with the Horrocksian method, and Jacques Cassini had a different approach to developing the "equation of the centre" (*GHA*, p. 201). As we shall see, Le Monnier was a key figure for French reception of *TMM*.

Chapter 6

Finding the Anomaly

This chapter surveys how astronomers in Restoration England dealt with the motion of a body obeying Kepler's second law. It deals with two fundamental concepts: "equation," as in for example the "Equation of the Centre," and "anomaly." Astronomers would proceed from what they called mean anomaly to the coequated (or, "true") anomaly, via the "Kepler equation." *TMM* instructs the reader to use tables at these crucial stages and that is all that we need to do. We are not obliged to go through the difficult stages then required to construct the tables, though it is appropriate that we should have some idea of their principles of composition.

I. The solar anomaly

Anomaly meant an angle, formed between two mean positions. It was a computational tool that could not be observed or measured directly. The mean anomaly was the angle between the apse line and a mean position in the orbit, at any given time. Whether it was measured from apogee or perigee was a matter of convention. To quote Curtis Wilson:

> The aphelion was taken as the zero of anomaly in planetary tables, and the apogee as the zero of anomaly in solar and lunar theory. (*GHA*, p. 275)

That became the agreed convention in the eighteenth-century. If however we go back to what Flamsteed wrote in 1680, that was not the case. For finding the Sun's anomaly, he explained:

> Subtract the Longitude of the Perihelion from the Mean Motion, the Residue is the Mean Anomaly." (*DOS*, p. 34)

It made sense to start from the Sun's perihelion, as it was close to the year's start in January. Tables gave their "Equation of the Centre" as a function of anomaly, over the range 0–180°. The anomaly values in these tables were not symmetrical, with their "equation" peaking at around 91° for the Sun and 94° for the Moon. Figure 6.1 (below) shows this in the case of the Moon, where M is the mean anomaly (here measured from apogee) and Θ is the Equation of Centre. Table 6.1 is a reproduction of a page from *DOS*, pointing out the maximum value reached at 94° anomaly.

The Earth's orbit around the Sun was an ellipse with "the Sun's Apogeum," as *TMM* called it, at one end: the point at which the Sun was furthest away, where it appeared to move most slowly. They are positioned in ecliptic longitude, for which reason we use the tropical year period in defining the mean sun, of 365.24 days. The motion of the Earth was treated in *TMM* as the motion of the Sun around the Earth.

Finding the mean anomaly was the first step in an arduous series of computations which the astronomer had to perform. The solar anomaly on the epoch date of December 31, 1680, for example, was the angular difference between *TMM*'s mean Sun at that time and perihelion position. This comes to 13° 11' (using the epoch value as given in Ch.4, Section II), which is the solar anomaly. This date is equivalent to January 10th, 1681 in New Style, i.e., Gregorian calendar, as there was a ten-day difference between the two systems in the seventeenth century. This date was 13 days after perihelion, reached on the morning of December 29th New Style, at $7\frac{1}{2}°$ Capricorn (For comparison, perihelion presently falls on January 3rd, at 12° Capricorn, so it has moved four and a half degrees in three centuries).

II Kepler Motion and the Equation of Centre

An "equation" then signified, to quote Curtis Wilson:

> the angle to be added or subtracted from a mean motion in order to "correct" it, that is, in order to obtain a theoretical position in agreement with the position observed. (*GHA*, p. 277)

TMM gave the Sun's "Equation of Centre," or "equation of orbit" as it was called a greatest magnitude of 1° 56' 20". The terminology derived from the old scheme of things, where the Sun had a circular orbit and Earth was not quite at the centre of that orbit, and the magnitude of its displacement from that point generated its "Equation of Centre." This equation was zero at the apses, growing to a maximum near the quadratures. It was subtracted while moving from aphelion to perihelion, and added during the other half of the year, since Earth's orbit is fastest at perihelion and slowest at aphelion.

Estimates of the maximal value of this solar equation had been shrinking ever since Tycho Brahe estimated it as two degrees. It is of interest to look at values cited by Curtis Wilson (*GHA*, pp. 168–191) as used by astronomers, comparing these with actual values for the period (the latter being derived from modern estimates of historic eccentricities):

Astronomer		Greatest Eqn. of Cent.	true value	Error
Brahe	1580s	2° 3′ 15″	1° 56′ 10″	7′
Horrocks	1638	1° 59′ 18″	1° 55′ 54″	3′ 22″
Cassini I	1660	1° 56′ 53″	1° 55′ 50″	1′ 37″
Flamsteed	1675	1° 54′ 13″	1° 55′ 49″	-1′ 36″
"	1679	1° 55′ 0″	1° 55′ 48″	-48″
"	1692	1° 56′ 20″	1° 55′ 45″	+35″

The last value was used in *TMM*, and in all the tables of "the Newtonians": Whiston (1715), Dunthorne (1739), Brent (1741), Leadbetter (1742), Le Monnier (1746) and Halley (1749). Indeed, I have only come across one compiler of astronomical tables in the first half of the eighteenth century in France or England who did not use that figure and that was Jacques Cassini (1740), sometimes called Cassini II. He used the more accurate value of 1° 55′ 50″.

The computations as then performed involved three stages:

1		2		3		4
mean anomaly	⇒	eccentric anomaly	⇒	coequated anomaly	:	apparent position

There was no simple means of moving from the mean anomaly to the "coequated" anomaly, so-called because an "equation" had been applied. It was done by using "Kepler's equation" to find what was called the eccentric anomaly. The Kepler equation is

$$\text{Mean anomaly} = E - e \sin E,$$

where the mean anomaly is measured from apogee, e is eccentricity and E is the eccentric anomaly. This was the angular position, as viewed from the *centre* of the ellipse, of a point moving in a circle circumscribing that ellipse. That point was collinear with the body on the elliptical orbit, the line being perpendicular to the major axis. Kepler's equation cannot be solved algebraically, so methods of approximate solution had to be developed and used (McNally 1974, p. 289). Through some means of solving "Kepler's equation" one obtained the eccentric anomaly from the mean (Colwell 1993, p. 10; Gaythorpe 1925, p. 864). Newton used a geometrical equivalent of what is called "Newton's method of approximation" to solve the Kepler equation (*PNPM* pp. 191-200; Whiteside *Math. Papers*, IV, 1971, p. 665; for a discussion of earlier approximate solutions by Ward, Bullialdus and Pagan, see Gregory 1702, pp. 386–8). Then the "coequated" anomaly had to be sought, which could be compared to the actual position (called, "apparent position') in the heavens. The goal of a theory was to minimise the difference between stages (3) and (4). What Newton had to say about moving between stages 1 and 2 appeared in the *Principia* and not in *TMM*, and need not concern us.

What the layman would call the true or actual position in the heavens, was and still is referred to by astronomers as the "apparent" position. This is because the "coequated" position used to be referred to as the "true" position. A lucid account of these terms has been given in *GHA* by Curtis Wilson (*GHA* Appendix, Gi-Gvi).

What is nowadays called the "equation of the centre" is not the angle as we have understood it above, but a formula which generates that angle. From values of eccentricity (*e*) and mean anomaly (*M*), it specifies that angle in radians by the following series expansion:

$$\Theta = (2e - e^3/4) \sin M - 5/4\, e^2 \sin 2M + 13/12\, e^3 \sin 3M + \dots$$

This is the modern expression for the difference between true and mean anomaly, i. e., stages (1) and (3) above. For the small eccentricity value of the Earth's orbit, the first two terms of this series expansion are generally adequate, which are:

$$\Theta = 2e \sin M - 1.25e^2 \sin 2M$$

These first two terms give the Equation of Centre within about half an arcminute for the Moon. To avoid confusion, we shall refer to the old meaning in upper case, as Equation of Centre, signifying an angle.

Let us now ask the historical question, when were tables of the Equation of Centre first prepared, as showed no significant deviation from the modern formula? As the science historians Thoren and Gingerich pointed out in 1974, the tables prepared by Flamsteed in 1679 for his *De Sphaera* were the first British tables to be computed from fully Keplerian principles. Gingerich and Welther found that the errors in the lunar Equation of Centre tables were up to 3" for minimum eccentricity and 10" for maximum eccentricity, concluding "On the face of it...in at least one important case Kepler's second law was being used in England before the publication of Newton's *Principia*" (Thoren 1974 pp. 243–256; Gingerich & Welther 1974, p. 258).

We may compare the lunar tables of Horrocks and Flamsteed in this respect, as the *DOS* Equation of Centre tables adopted the same maximum and minimum eccentricity values as had those earlier-published by

A TABLE of the Equations of the Moons Center.

Subtract.

Mean Anomaly	Sign 2. LeaftExc 43619	Middle 55327	Greateft. 66854	Sign 3. LeaftExc 43619	Middle 55327	Greateft. 66854	Mean Anomaly
	° ′ ″	° ′ ″	° ′ ″	° ′ ″	° ′ ″	° ′ ″	
0	4 12 40	5 17 27	6 21 18	4 59 30	6 18 59	7 38 17	30
1	4 15 18	5 20 54	6 25 32	4 59 48	6 19 23	7 38 52	29
2	4 17 56	5 24 17	6 29 39	4 59 56	6 19 40	7 39 20	28
3	4 20 28	5 27 35	6 33 40	4 59 59	6 19 50	7 39 40	27
4	4 23 00	5 30 47	6 37 36	4 59 58	6 19 54	7 39 51	26
5	4 25 24	5 33 53	6 41 25	4 59 49	6 19 51	7 39 53	25
6	4 27 44	5 36 54	6 45 08	4 59 36	6 19 40	7 39 47	24
7	4 30 00	5 39 49	6 48 44	4 59 20	6 19 23	7 39 33	23
8	4 32 12	5 42 39	6 52 14	4 58 53	6 18 57	7 39 09	22
9	4 34 19	5 45 24	6 55 36	4 58 24	6 18 25	7 38 37	21
10	4 36 21	5 48 02	6 58 52	4 57 48	6 17 46	7 37 58	20
11	4 38 18	5 50 35	7 02 01	4 57 06	6 17 00	7 37 09	19
12	4 40 12	5 53 02	7 05 03	4 56 19	6 16 08	7 36 12	18
13	4 41 58	5 55 22	7 07 57	4 55 27	6 15 08	7 35 06	17
14	4 43 41	5 57 36	7 10 45	4 54 30	6 14 00	7 33 52	16
15	4 45 19	5 59 44	7 13 25	4 53 27	6 12 46	7 32 29	15
16	4 46 53	6 01 46	7 15 58	4 52 19	6 11 25	7 30 57	14
17	4 48 22	6 03 42	7 18 24	4 51 03	6 09 56	7 29 17	13
18	4 49 44	6 05 31	7 20 42	4 49 45	6 08 20	7 27 28	12
19	4 51 02	6 07 15	7 22 53	4 48 21	6 06 37	7 25 30	11
20	4 52 15	6 08 52	7 24 56	4 46 51	6 04 48	7 23 23	10
21	4 53 22	6 10 23	7 26 52	4 45 16	6 02 51	7 21 08	9
22	4 54 23	6 11 46	7 28 39	4 43 34	6 00 48	7 18 44	8
23	4 55 20	6 13 03	7 30 20	4 41 44	5 58 37	7 16 12	7
24	4 56 12	6 14 14	7 31 52	4 39 56	5 56 19	7 13 30	6
25	4 56 59	6 15 19	7 33 16	4 37 58	5 53 54	7 10 40	5
26	4 57 38	6 16 17	7 34 32	4 35 56	5 51 23	7 07 42	4
27	4 58 14	6 17 08	7 35 40	4 33 47	5 48 43	7 04 35	3
28	4 58 45	6 17 52	7 36 41	4 31 33	5 45 57	7 01 19	2
29	4 59 10	6 18 29	7 37 34	4 29 13	5 43 04	6 57 55	1
30	4 59 30	6 18 59	7 38 17	4 26 40	5 40 05	6 54 23	0
	Sign 9.			Sign 8.			

Add.

Table 6.1: Lunar Equation of Centre Tables in *DOS*, for three different eccentricities, showing two columns spanning 60°-90° and 90°-120°, for three different eccentricities. The values peak around 94° of anomaly, at almost 6° 20′ (underline added).

Horrocks. We compare the errors, for anomaly values from 60°, 75° and 90°, in these two sets of tables, by subtracting from them the modern equation of centre values. Flamsteed's anomaly values are shown above in Table 6.1. The values whose computed "errors", i.e., deviation from Kepler-motion are shown, ranged from 4°12' to 7°38'.

Error-values at max. & min. eccentricity positions, in arcseconds:

			60°anomaly	75°anomaly	90°anomaly
Horrocks	1673	Max/Min:	13″/ 2″	13″/ 2″	12″/-2″
DOS	1681	Max/Min:	0″/ 6″	4″/ 1″	5″/-1″

It is clear that the *DOS* accuracy has somewhat improved upon its predecessor. Flamsteed and his assistant Mr Hodgson accomplished these computations. Surprisingly, the modern series expansion concurs within *one or two seconds of arc* with the Equation of Centre tables of Flamsteed, indicating that by 1681 at least one astronomer had effectively solved the problem of computing elliptic, Kepler motion.

If anything, *DOS*'s lunar Equation of Centre tables had a slightly higher level of accuracy than claimed by Thoren & Gingerich. Flamsteed's computation methods used the first and second laws of Kepler, at a time when Newton himself had not then used Kepler's second law in any astronomical context (Cohen 1980, p. 250). Flamsteed's assistant Mr Hodgson recalled:

> Mr Flamsteed, under whom I had the happiness of my education, was pleased to set me upon computing his lunar tables, under his direction; when I computed the tables of central equations of the moon after the Keplerian method, which had never been done before. (Hodgson 1750, Introduction; Baily 1837, p. 704).

By whatever means, he has somewhat improved upon Horrocks's Equation of Centre table as was published a decade earlier. For comparison, Table 6.1 shows *DOS*'s middle table of three for Equation of Centre, for 60°–120° anomaly.

As an example of how an astronomer would use tables, let us seek for a solution whose solar anomaly value (as found earlier) was 13°11.' For this given value, we consult tables for the solar Equation of Centre. Options were available:

	DOS (1681)	Dunthorne (1739)	Cassini (1740)
13°	25′ 21″	25′ 38″	25′ 33″
14°	27′ 16″	27′ 34″	27′ 29″
Max.Eqn.(91°)	1°55′ 0″	1°56′ 19″	1°55′ 50″

The astronomer had to interpolate between the values given. Dunthorne used Newton's eccentricity value. For an angle of 13° 11′, Dunthorne's tables give 25′ 59″. The modern formula gives the identical value. The *DOS* tables gave a slightly smaller value, having been composed before Flamsteed had reached his final and more exact value for the earth orbit's eccentricity. This equation brings us to within half an arc minute of the Sun's actual position. Possibly the solar position was departing from Kepler-motion by that amount.

TMM specifies for an elliptic orbit both the eccentricity and maximal Equation of Centre, and the the above equation relates these together. In the case of the solar orbit, *TMM* specifies eccentricity as $16\frac{11}{12}$ parts in 1000, and the maximum Equation of Centre as 1° 56′ 20″. The anomaly value which generates the maximal Equation of Centre is $M = 91°$, i.e., solar tables will give the greatest "equation" at 91°. For the lunar tables the maximal value arises at or near to 94°. Inserting 91° into the above equation for M links *TMM*'s two solar eccentricity parameters together within one second of arc. Although *TMM* nowhere mentions elliptical orbits, this identity establishes that a Keplerian definition of eccentricity is being used.

III. The Lunar Equation of Centre

The lunar "equation of centre" was a more complicated affair, varying not only with its mean position in orbit, but also with a half-yearly cycle. The Horrocksian theory used the altering eccentricity of the lunar orbit to modulate the amplitude of the "equation of centre." Here we merely introduce the subject, prior to a full account in the next chapter.

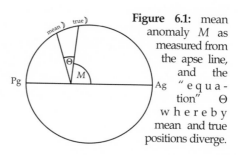

Figure 6.1: mean anomaly M as measured from the apse line, and the "equa-tion" Θ whereby mean and true positions diverge.

A mean moon has uniform angular velocity in ecliptic longitude, revolving once per tropical month. To obtain the lunar anomaly one subtracted the "mean apogee" position therefrom. In defining mean anomaly as an angle measured from the mean apogee position, we are treating the motion of the apse as a continuous function, whereas in fact the apogee and perigee positions only exist at discrete positions once a month. The concept is a mathematical abstraction, defined as an angle between two points in uniform motion, in the same direction around the ecliptic. These intersect once per 27.554 days, the period of the anomalistic or apogee-perigee cycle.

IV Finding the Prostaphaeresis

The amplitude of the Equation of Centre varied in accordance with what *TMM* called the "Annual Argument." This was the term employed by Horrocks, who took it from Kepler (Wilson 1987, p. 81). Chapter Two described its $6\frac{1}{2}$ month period between successive alignments of the syzygy and apse lines. As syzygy is the line joining full and new Moon, this conjunction happens between Sun and the mean apse in zodiac longitude, twice per "Horroxian year" of 411 days. The angle between Sun and mean apse will here be termed the "Horrox angle" (Φ). One may prefer not to use the *TMM* term "Annual Argument" of the apogee, as the period is not really annual. It appears in Figure 6.2 as 46°.

Over this interval, *TMM* specified that the lunar Equation of Centre's maximum value should range between 7° 39′ 30″ and 4° 57′ 56″. This "Equation of the Moon's centre" was sometimes called the *Prostaphaeresis,* a function still harder to find than it was to pronounce.

The first step in determining this "equation" is to find the Horrox angle, Φ. As a first approximation one could view the peak value of this "Equation of the orbit" as a cosine function, maximal at zero degrees when the Sun is in conjunction with the lunar apse and minimal when the two axes are perpendicular, i.e., a cosine 2Φ function having two cycles per Horroxian year. Its magnitude would then be of the form:

$$\text{Equation of Centre} \approx \alpha + \beta \cos 2\Phi$$
$$\text{with a maximum value of} \quad 7° 40', \text{ when } \Phi \text{ is zero,}$$
$$\text{and a minimum of} \quad 4° 58', \text{ when } \Phi = 90°,$$

as *TMM* requires. This is merely intended to indicate how the functions are linked. Thus, the greatest "equation of the centre" is half as much again as its least value, whereby the second or Kepler-motion moon comes to differ by a greater amount from the mean moon. *TMM* develops a more complex kinematic/geometrical approach described in the next chapter.

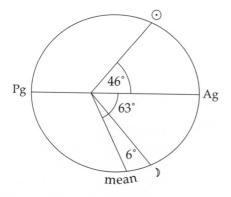

Figure 6.2: Mean positions on epoch date of December 31st, 1680.

Considering the position shown in Figure 6.2, on the 1680 epoch date, the mean moon was at 2° of Libra and the mean apogee at 5° Sagittarius, giving a mean anomaly of -63°. The Moon was then ahead of its mean position, having travelled faster while near perigee. The Sun-apse "Horrox angle" was 46°, and is here generating an equation of centre equal to six degrees.

In this chapter we have, as a first step, performed some computations to give the Sun's position within half an arcminute and the Moon's within half a degree. This may give us some respect for the difficulties involved, and a notion of how to apply an "equation."

Chapter 7

The Horrox-Wheel in Motion

Horrox had left no description of the theory itself, but Crabtree was helped in his reconstruction by rough diagrams drawn on loose papers... (Forbes 1975, p. 63).

TMM embodies a developed version of Jeremiah Horrocks' lunar theory, what one might call Newton's interpretation of Halley's variation of Flamsteed's version of Crabtree's account of Horrocks's lunar theory. Had Horrocks lived beyond the brief span of twenty-two years, he might have described his theory more fully; and yet, even in the incomplete state in which he left it, it was in Flamsteed's view the greatest of his achievements.

Curtis Wilson has described how the theory began to dawn on Horrocks in January 1637 (Wilson 1989, p. 86), and had been formed by December 1638, when he prepared "the new calculus of the Moon" and sent it to William Crabtree. Figure 7.2 is a diagram drawn by Crabtree showing how the process unfolds through a full thirteen-month cycle. This diagram first appeared in Horrocks' letter to Crabtree dated 20 December 1638 (Horrocks 1673 p. 471), and was then sent from Crabtree to William Gascoigne in June 1642. These were the three north-countrymen who initiated the tradition of British astronomy, whose work became known via Flamsteed, as he moved from Derby to London in the early 1670s.

Flamsteed became the chief exponent of the Horroxian theory, such that astronomers knew it largely as Flamsteed's development thereof, as presented in his epilogue to *Jeremiae Horoccii,...Opera Posthuma* published by John Wallis in 1673. A succinct version thereof was given in a letter of Flamsteed's to Newton (*Correspondence*, Vol. IV, p. 27). Whiston referred to it as "Mr *Horrox's* Lunar Hypothesis, as cultivated and explained by Mr Flamsteed" (Whiston 1726, p. 104). Earlier, Streete had propounded a version of Horrocks' lunar theory in his his *Planetary System* of 1674, a development of the theory he to some extent presented in his *Carolina* (Kelly 1991, p. 52).

I. A Variable Eccentricity

The evection, taken in conjunction with elliptic inequality, has the effect of rendering the eccentricity of the moon's orbit variable. (Gaythorpe 1871, p. 70)

Figure 7.1: Diagram in
Horrocks's notebook for
computing semi-annual
variations in lunar eccentricity and apse (his
'Philosophical Exercises')

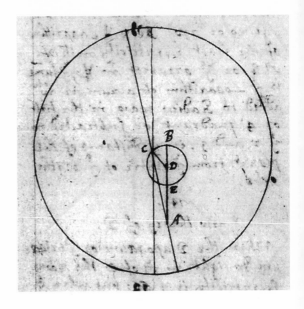

Figure 7.2: Diagram drawn by
William Crabtree to illustrate
Horrocks's theory of the
evection (in his letter
to Gascoigne of
June or July
1642)

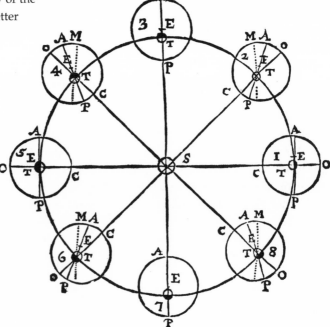

In dealing with the Horrocks model, we started by recalling Hipparchus' concept of eccentricity, with its image of circular motion about an epicentre, where Earth's displacement from that epicentre *was* the eccentricity. Then, in the seventeenth century, eccentricity became the distance between focus and centre of an ellipse, where the circumscribing circle had unit radius. Whiston's textbook had a section, "To Determine the Earth's Eccentricity," which explained how eccentricity was "to be reckoned from Focus to Center." We may note that diagrams of the Horrocks theory always showed circular orbits with displaced centres, as indeed Kepler had done: "Kepler neglects the elliptical shape of the orbit in computing the evection.... The error could not be very large, since the moon's orbit is constricted by only about $e^2 = 0.002$ of its radius." (B. Stephenson 1987, pp. 182, 186).

In the Horroxian model, the apse line and eccentricity co-varied, by a similar amount and 180° out of phase. The Crabtree diagram (Figure 7.2) displays the twin features of apse line oscillation and eccentricity variation, out of phase with each other. In each such period the Sun met the apse line, shown in steps 3 and 7. These conjunctions coincided with maximum lunar eccentricity, while in steps 1 and 5 of the diagram it becomes minimum. William Whiston, Newton's successor at the Lucasian chair at Cambridge, reproduced this Crabtree diagram in his textbook (based on lectures given in 1703), and commented:

> It is to be noted that the Eccentricity of the Lunar Orbit is mutable; and that the same, in the Conjunction and Opposition of the Apogee, is One and a Half of the Eccentricity, which is in the quadratures. So that TE the Distance between the Focus and the Center of the Ellipsis, in the position of her Orbit, marked [3] is One and a Half of the same Distance in the Position marked [5]. (Whiston 1726, p. 104)

In this crucial diagram, the primary motion of the apse line has been subtracted out, so that it only depicts the secondary oscillation. It does *not* depict rotation in sidereal space: after one revolution through its eight stages, taking 411 days, the apse line will have revolved 46°, whereas it is represented by a vertical line in each phase. At each step, the dotted line MT representing the mean apse remains parallel to the vertical axis of the diagram. The apse "equation" is the angle between the "equated" apse line AP (apogee-perigee) and the mean apse, and this peaks at the four octant positions, i. e., in between the quadratures. We also note that the eccentricity ET is inscribed upon the equated apse line AP.

As Whiteside in his tercentenary essay of 1976 affirmed, the young (22-year-old) Horrocks had constructed the following hard-to-picture edifice:

> The theory of the moon's motion thereby subsumed [i.e., the Horrocks theory], by which (on conflating the Ptolemaic equation of excentre—essentially our modern elliptical inequality —and the evection from this) the lunar orbit is taken to be basically a Keplerian ellipse of periodically varying eccentricity with a corresponding fluctuation to and fro in the mean secular advance of its line of apsides, and further adjusted *in fine* by—in longitude—Brahe's twin inequalities of variation and the annual equation. (Whiteside 1976, p. 318)

Figure 7.1 shows the Horrocks diagram, as it appears in his unpublished *Philosophical Exercises* (This notebook is stored together with Flamsteed documents in the R.G.O. archives at Cambridge University Library, RGO 1.68B; *GHA* p. 197). Curtis Wilson's researches showed that Horrocks derived the diagram from van Lansberge's *Theoricae motuum coelestium,* but that he altered the theory involved, so that from the "very inaccurate" Lansberge model, he constructed what remained for almost a century the finest available. I remained in the dark as to how the Horrocks model functioned, until I started to follow carefully the instructions given in *TMM*, its diagram being given in Figure 7.3. The two diagrams are basically equivalent, though separated in time by six decades.

II. TMM's Diagram

In the *TMM* diagram, an immobile Earth is positioned at *T*, around which the Sun *S* revolves yearly, and an immobile mean apse line *TB*, around which we could picture the stars revolving every nine years. That is the frame of reference in space and time required by the model. *TF* is the apse line varying by its second equation (its first equation, the annual, is not here represented). In Horrocks' version, Figure 7.1, a variable centre *C* to the lunar orbit is defined. Likewise, in the *TMM* diagram, Figure 7.3, the points *C* and *F* represent the mean and first-"equated" centres of the ellipse of the lunar orbit.

According to Newton and Halley, the eccentricity was represented by the line *TF* in Figure 7.3, whereas according to Flamsteed it was represented by the projection of that line onto *TB*. These have the same maximum and minimum values, *TB* and *TA* respectively, but different mean values. Authorities have agreed in taking the former as correct, i.e., as equivalent to the modern equations for the evection inequality, but have as we shall see disagreed over which was employed by Horrocks.

In figure 7.3, *STA* is the Horrox-angle Φ, and *FCB* is defined as being of double its magnitude, i.e., 2Φ. Unfortunately, *TMM*'s diagram is badly drawn and wrongly makes *TS* and *CF* appear as parallel. For *TMM*'s eccentricity equation, we require the length of *TF* in terms of Φ. *TC* represents the mean eccentricity namely 0.055050 and *CF* is half the difference between maximum and minimum eccentricity, namely 0.011731. The important ratio of 21.3% as its fluctuation (Ch. 2, IV) is here given geometrically by *CF/CT*. Applying the cosine formula in triangle *FTC* gives:

$$TF^2 = TC^2 + FC^2 - 2FC.TC \cos FCT$$

$$TF = TC\sqrt{1 + FC^2/TC^2 - 2FC/TC \cos FCT}$$

$$= 0.05505\sqrt{1 + 0.2131^2 + 2 \times 0.2131 \cos 2\Phi}$$

or
$$TF = 0.05505\sqrt{1.0454 + 0.4262 \cos 2\Phi} \qquad (7.1)$$

That is our equation representing the Newtonian instructions in *TMM* for finding the eccentricity. Using the same terms, but with an additional perpendicular dropped onto the mean apse as shown in figure 7.4a, Flamsteed's version as given above was much simpler:

$$\text{Eccentricity} = TD = TC + CD \qquad \text{(see Figure 7.4a)}$$

$$= 0.05505 + 0.01173 \cos 2\Phi \qquad (7.2)$$

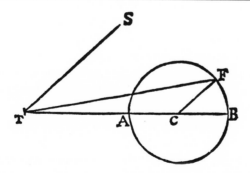

Figure 7.3: *TMM*'s diagram of the Horrox-wheel, where angle *FTB* is the apse equation δ, *STB* the Sun-mean apse angle Φ and *FCB* is 2Φ, though not well drawn to scale (*ST* should be parallel to *AF*). *TC* is mean eccentricity, while *TA* and *TB* are its minimum and maximum values respectively.

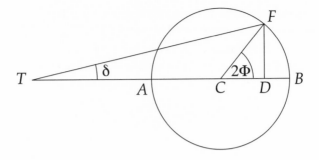

Figure 7.4a: Horrox-wheel diagram showing *TD* as Flamsteed's version of eccentricity and *TF* as Newton's. *TB* is the mean apse line and *F* is the centre of the lunar orbit.

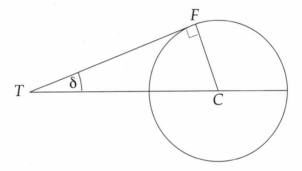

Figure 7.4b: the condition for maximum δ, where sin δₘₐₓ = *FC/TC*. *TF* is a tangent to the circle.

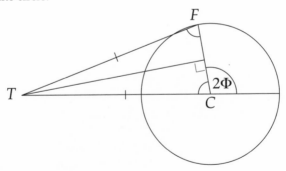

Figure 7.4c: *TMM*'s mean eccentricity position: *TF = TC*, where *TF* is not tangential to the circle, then cos *FCT = CM/CT = FC/2TC*.

The two equations differed considerably in their effect.

To obtain "Ye Equation of Ye Apogee" as Flamsteed called it, or the "second Equation of the Moon's apogee," as *TMM* called it, we require the angle *FTC*. Following the example of Curtis Wilson in *GHA* we take this as δ, then applying the sine formula to the same triangle gives

$$\sin \delta = (0.011732 \sin 2\Phi)/e,$$

$$= \sin 2\Phi/85.24\ e \tag{7.3}$$

Flamsteed's version took a smaller value for the radius of the Horrox-wheel (see Table 7.2 below). In the next chapter, these procedures for obtaining the length *TF* , the "equation" of eccentricity, and the angle δ, *TMM*'s second apse "equation," will be referred to as the functions *f* and *g* respectively. These functions are constructions from the triangle *TFC* using the sine and cosine formulae.

III. Offbeat Octants

These Newtonian functions are asymmetric about the octants of Figure 7.2. As appears in figure 7.4b, the angle δ maximises when *TF* is a tangent to the circle, i.e., when *TFC* = 90°. Then,

$$\sin \delta_{max} = FC/TC = 0.2131 \tag{7.4}$$

so $\qquad\qquad \delta_{max} = 12° 18' 15''.$

This maximal value arises when the Horrox angle is given by $\cos(180 - 2\Phi) = 0.2131$, or $\Phi = 51°$. Whether or not this asymmetry has any astronomical significance, the tables which follow *TMM*'s instructions all have their apse equation peaking at 51° of the "Annual Argument." Table 7.1 shows the table which Newton sent to Flamsteed in April of 1695, which maximises at a Horrox-angle of 51°.

The three columns of this Table each span 30°, so that for example the middle column's peak value at 21° represents 51° of the "Annual Argument," i.e., the Sun-apse angle. The peak apse-equation value of 12° 10' exceeds that given in *DOS* (11° 47') but is smaller than *TMM*'s value of 12° 15', demonstrating that *TMM* was not composed in this period. Adjacent to this column in the Table are the eccentricity values, where the mean value (55050 parts in 10^6) appears as just after 48° of the "Annual Argument."

Table 7.1: Table sent by Newton to Flamsteed on 23 April 1695, with three columns 0°–30°, 30°–60° and 60°–90°, for the Horrox angle ("Annual Argument"); showing mean eccentricity of 55050 generated by such an angle of 48–49°, while maximal apogee equation (12° 10′ 25″) falls at 51°.

The Equations of the Moons Apoge & the Excentricities of her Orbit
in such parts as the radius is 1000000.

		Add the Equations of the Apoge						
Annual Argument	Sign 0 6	Excentr.	Sign 1 7	Excentr	Sign 2 8	Excentri	Annual Argument	
	° ′ ″	parts	° ′ ″	parts	° ′ ″	parts		
0	0 0 0	66850	9 22.50	61855	11.32.17	50406	30	
1	0.20.54	66845	9.36.57	61537	11.22.59	50022	29	
2	0.41.46	66827	9.50.31	61211	11.12.37	49645	28	
3	1. 2.38	66798	10. 3.40	60878	11. 1.10	49274	27	
4	1.23.27	66757	10.16.14	60438	10.48.39	48908	26	
5	1.44.12	66705	10.28.17	60192	10.35. 2	48551	25	
6	2. 4.54	66638	10.39.47	59838	10.20.21	48201	24	
7	2.25.31	66562	10.50.41	59479	10. 4.36	47859	23	
8	2.46. 0	66475	11. 0.58	59113	9.47.47	47527	22	calculo proprio
9	3. 6.24	66375	11.10.40	58742	9.29.55	47204	21	correcti sequentes
10	3 26.41	66265	11.19.42	58366	9.10.59	46891	20	46891 numeri
11	3.46.50	66146	11.28. 5	57986	8.50.58	45266	19	46588 JF
12	4. 6.48	66012	11 35.46	57600	8.29.57	45040	18	46298 Maij 4 ♄
13	4.26.37	65870	11.42.44	57211	8. 7.57	44829	17	46019 1695
14	4.46.15	65716	11.48.58	56819	7.44.58	44633	16	45753
15	5. 5.41	65549	11.54.27	56422	7.21. 1	44452	15	45500
16	5.24.55	65373	11.59.11	56023	6.56. 8	44287	14	45260
17	5.43.53	65185	12. 3. 6	55622	6.30.23	44138	13	45034
18	6. 2.38	64988	12. 6.12	55218	6. 3.49	44824	12	44823⅓
19	6.21. 9	64779	12. 8.28	54814	5.36.28	44628	11	
20	6.39.22	64562	12. 9.53	54408	5. 8.22	44447	10	Mean Value
21	6.57.20	64343	12.10.25	54001	4.39.34	44283	9	55050
22	7.14.55	64094	12.10. 1	53595	4.10. 8	44134	8	
23	7.32.14	63847	12. 8.43	53190	3.40.10	44003	7	
24	7.49.11	63590	12. 6.28	52784	3. 9.41	43888	6	
25	8. 5.47	63323	12. 3.16	52381	2.38.45	43789	5	
26	8.22. 0	63046	11.59. 6	51980	2. 7.27	43709	4	
27	8.37 51	62761	11.53.58	51581	1.35.51	43647	3	
28	8.53.17	62467	11.47.45	51185	1. 4. 2	43602	2	
29	9. 8.17	62165	11.42.14	50794	0.32. 3	43575	1	
30	9.22.56	61855	11.32.17	50406	0. 0. 0	43566	0	
	5 Sign 11		4 Sign 10		3 Sign 9			
	Substract the Equations of the Apoge							

$57·02 = 15′.44″$

Returning to Crabtree's diagram (Figure 7.2), eccentricity there appears as reaching its mean value at the octants, as indeed it did in Flamsteed's version, given by equation 7.2. For the Newtonian eccentricity however (TF in Figure 7.4c), its mean value occurs when $TF = TC$, and then cos $FCT = \cos(180-2\Phi) = CM/CT = 0.2131/2$, giving $\Phi = 48°$, so that "Newtonian" tables for eccentricity reached their mean value at 48°. Again, I am not clear as to whether this has an astronomical meaning. It conferred upon one pair of octants (moving from conjunction to square) a different form of motion from the other two (from square to conjunction).

Thus, the "Horroxian" table of lunar eccentricity which Flamsteed published in 1673 (*Lunares Numeri Ad Novam Lunae Theoriam*, p. 480) resembles the equivalent table in *DOS* in having a mean value at 45° of the Horrox angle, and so being symmetrical about the octants. This difference offers a simple and distinctive fingerprint for recognising who in the eighteenth-century was using Newton's version of the Horroxian mechanism.

IV. Halley's Contribution

Halley afterwards made a slight alteration; but hardly, I think, enough to justify Newton's assertion. (Whewell 1837, Vol. 1 p. 466)

An adjustment to the Horrocksian model was recommended by Halley to Newton, which the latter regarded as quite valuable. Confusion has arisen over this matter, for the resolution of which we need to review the development of what was regarded as Horrocks's lunar theory. It took place in four stages:

I	II	III	IV
Horrocks (1638) notes	Crabtree (1642) Letters	Flamsteed (1673) *Opera Omnia*	Flamsteed (1681) *DOS*

Streete (1674)   Whiston (1717-26)

The first coherent account of Britain's first lunar theory emerged from Salford, now a suburb of Manchester, in June of 1642, as penned by William Crabtree to Gascoigne from notes left by Horrocks. They had both been his colleagues. Crabtree cast the new theory into seven steps, the third of which is here of interest to us. In the 1673 publication of Horrocks's *Opera*, Flamsteed "polished" Horrocks's method, to use his expression, chiefly by inserting his own Equation of Time in place of the imaginary Keplerian "equation of physical parts" in which Earth's rotation rate altered through

the year. In 1681, Flamsteed adjusted some of the mean motions as compared with his earlier 1673 statement, but otherwise left his method
unchanged. Whiston merely repeated the *DOS* procedure, presenting it to
his students in the early decades of the new century as the best lunar theory
available. As mentioned earlier, Streete was one of the first astronomers to
use the new Horroxian theory.

Crabtree's letter was reprinted in Flamsteed's 1673 collection of
Horrocks's *Opera*. Its third step used logarithms to specify the manner in
which eccentricity varied in the new theory:

> 3....Duplicetur Argumentum annum, & duplicati Co-sinui
> addatur 3,065206 (Logar. numeri 1162, semi-differentiae inter
> mediam & extremam Excentricitatem) prodibit Logarithmus
> numeri addendi Excentricitatem mediae 5524, si duplum Arg.
> annui fit in 4° vel 1° quadrantibus, alias subtrahendi, &
> habetur Lunae excentricitatis vera. (Horrocks 1673, p. 469)

This text describes the addition of mean eccentricity (the line *TC* in Figure
7.4a), here given the value of 5524, and what was earlier described as *FC* cos
2Φ (see equation 7.2); the radius of the Horrox-wheel having the magnitude
1162, described by Crabtree as half the difference between maximum and
minimum eccentricity. The cosine term is of twice the "annual argument" as
it was called. Terms are converted to logarithms and instructions given as
to whether the *FC* cos2 Φ term is added or subtracted quadrant by by
quadrant—done automatically by our cosine function. Horrocks's method
projected the rotating radius vector *FC* onto the mean apse line *TC* to define
the altering eccentricity.

Horrocks's original words on the subject are to be found in his own notebook, *Philosophical Exercises* (RGO 1.68 B, Second Part, section 19), entitled,
"A New Theory of the Moon." Let us emphasise that no science historian
prior to Curtis Wilson has alluded to or even apparently noticed this notebook. Earlier writers on Horrocks (Whewell, Gaythorpe, Whiteside) did not
mention this source. It was written in English:

> To the sine of the remainder adde 306446 (the logarithme of
> 1160, or halfe the difference of the greatest and least
> eccentricity of the Moon) so have you the logarithm of a num-
> ber to be added to 5493 (the middle eccentricity)...so have you
> the moons eccentricity."

What Horrocks meant by "the remainder" pertained to double the "Sun's
distance from ye moon's apogaeum or perigaum." He was then using we
may note, a slightly lower mean eccentricity value.

Flamsteed adopted the account in Crabtree's letter, whereby eccentricity was represented by the projection of the line *TF* (see Figure 7.4a) onto *TB*. In the early 1690s, Halley came to disagree with this, averring that the eccentricity should rather be represented by the line *TF* itself. We lack any statement by Halley on this matter, which is regrettable. Flamsteed reported it to Newton, evidently puzzled, in his letter of 11 October, and then added, referring to a diagram: "But he [Halley] affirms that not *Cx* but *CI* is the excentricity in this & all other cases" (corresponding to lines *TD* and *TF* respectively in 7.4a). Newton's reply a week later was,

> By your observations I find it to be a very good correction. I reckoned it a secret which he [Halley] had entrusted with me; and therefore never spake of it till now. (*Correspondence* IV, p. 34, letter of 24 October, 1694)

The *Principia* of 1713 gave this curious account of events:

> Our countryman Horrox, was the first who advanced the theory of the moon's moving in an ellipse, about the earth placed at one focus. Dr Halley improved the notion, by putting the centre of the ellipse in an epicycle whose centre is uniformly revolved about the earth; and from the motion of the epicycle the mentioned inequalities in the progress and regress of the apogee, and in the quantity of eccentricity, do arise." (*PNPM* p. 661)

The first sentence describes what would widely be viewed as the achievement of Johann Kepler, attributing it to Jeremiah Horrocks, and the second seems to describe the achievement of Jeremiah Horrocks, attributing it to Halley. Kepler's lunar theory was, it must be said, tentative, and he did not have the same confidence in applying an elliptical orbit to it as he had done with the planets.

It was fairly well known that Kepler had applied his laws to the lunar orbit but had not thereby made much progress (Wilson 1987, p. 80). An article by S. B. Gaythorpe, F.R.A.S. in 1957 on "Jeremiah Horrocks and his 'New Theory of the Moon'" commented upon the *Principia's* text, that it

> Does not indeed seem a particularly striking claim to fame, but the sentence implies more than it immediately conveys. Before Horrox no one had attempted to take an ellipse as the basis, so to speak, of the Moon's path, on account of the number, size, and rapid variation of the periodic inequalities involved, and the difficulty of combining them with other than circular motion." (Gaythorpe 1957, p. 134)

As well as endorsing the *Principia*'s version of Horrocks originality, Gaythorpe concluded that Horrocks had omitted a certain factor (sec δ) in the eccentricity formula, and thereby "he lost the honour which Newton gave instead to Halley." (Gaythorpe 1957, p. 137) Sec δ, or cos δ, is the factor by which the two recipes disagree (see Figure 7.4a).

A different viewpoint appeared in Whiteside's essay on the subject, affirming that Halley had not made any innovation, but had merely adopted Horrocks's method:

> It was Flamsteed's understanding (founded on a passage in Horrocks where he himself uses this simplification for ease of calculation) that the eccentricity of the lunar ellipse is not—as Horrocks himself indubitably took it to be in his basic precepts —*TF* [in Figure 7.4a] but the projection of this on to the mean apsis-line. (Whiteside 1975, p. 325, note 10)

Whiteside gave as his authority Gaythorpe (1956, p. 137). As we have seen, however, this is not Gaythorpe's view. Curtis Wilson in the *GHA* endorsed the Whiteside view:

> On one point of interpretation Flamsteed went astray, thereby deeply changing the structure of Horrocks's theory: in place of the varying eccentricities intended by Horrocks, he used their projections onto the mean line of apsides, so diminishing their values by the factor cos δ. The mistake was later perceived and corrected by Halley. (*GHA*, p. 198)

What Halley imparted to Newton as a secret, and to Flamsteed as his own idea, is here viewed as merely a reversion to the earlier model. (See also *Correspondence*, vol. IV, p. 32, note 6, discussing the above-mentioned letter of 11 October). Admittedly, Horrocks's figure (7.1) does not suggest any projection of the point *C* onto the mean apse line *AD*, and Crabtree's diagram (Figure 7.2) unequivocally measures eccentricity along the equated apse line, as was no doubt the basis for Whiteside's assertion. All one can say with confidence, is that all written accounts of the method prior to *TMM* measured eccentricity along the mean apse.

We may here consider what was then meant by the term "Horrocksian." In 1710 a Mr Cressner published in the *Philosophical Transactions* a comparison of two different longitude computations, one of which he called Newtonian and the other Horroxian (*Phil. Trans.* 27, pp. 16–19). The two methods will be compared in Chapter Ten. The latter turned out to be the account given in William Whiston's *Praelectiones astronomicae* (1707) of

Flamsteed's 1681 *DOS* procedure.

On the Forbes-Gaythorpe view, Flamsteed's eccentricity procedure was that of Horrocks, and Halley's proposal was an innovation. Could Flamsteed have mistaken the theory which he brought from the North Country and ushered into the light of day? A later chapter will look at the question of how much difference was made by this adjustment, whether it was a "very good correction" (Newton) or but a "slight alteration" (Whewell).

Forbes affirmed that a geometrical construction for the altering eccentricity was supplied "for the first time" by Flamsteed in his epilogue to the Horrox *Opera Posthuma* of 1673 (Forbes 1975, pp. 63–67). Figure 7.1, showing Horrocks' (unpublished) eccentricity equation and deferent-wheel, from Horrocks's "Philosophical Exercises" adjacent to the passage above-quoted, was unknown to him.

V. **Linkage of** *e* **and** δ

How well do the apse and eccentricity equations tie up together, in the several versions of the Horrocksian theory? Flamsteed's version of the model in *DOS* effectively had two deferent wheels concentric upon the centre of the lunar orbit, that for the apse equation being 2.9% smaller than that for the eccentricity function, as was also the case for Horrocks' theory. Not prior to *TMM* did their magnitudes coincide. *TMM*'s instructions seem to indicate that the maximal value of the Equation of the Apogee is a simple function of the varying eccentricity, as in the above equation (7.4), however, its figures do not quite bear this out:

	Eccentricity	Eqn.of Apse line	Predict.Eqn.
Horrocks	0.05524 ± 0.01162	11°47′22″	12° 8′35″
Flamsteed		11°47′22″	
Newton (1695)		12°10′25″	
(1702)	0.055050 ± 0.011732	12°15′ 4″	12°18′15″
" (1713)	"	"	12°18′.

Table 7.2: Maximal "equations" of eccentricity and apse in Horroxian models, 1642–1713, with predicted values for apse equation derived from Horrocks-wheel geometry (equation 7.4).

In his letter of June 21 1642 to Gascoigne, William Crabtree cited Horrocks's value of the apse equation quoted above, as was in turn cited by

Gaythorpe (1956, p. 137). However, Horrocks' *Philosophical Exercises* note-book stated "The greatest *aequatio apogai* is 12°30." (RGO 1.68B, 17) In 1675, Flamsteed in a letter to the *Philosophical Transactions* averred that: "I find by Mr *Horrockses* papers, that he used at first 12° precise, but upon farther experience diminished it to 11°48.' (*Phil. Trans.* 1675, 10, pp. 369-70.)

Not prior to the *Principia's* second edition were the values interlinked in accord with *TMM's* geometrical model. It thus appears that the form of the Horrocks model, above described, was originated by Newton. Dunthorne in his *Practical Astronomy of the Moon* faithfully reproduced the instructions of *TMM* in drawing up tables, etc, but gave an Equation of the Apogee as 12°18'15", and it now becomes clear that Dunthorne has simply calculated its value from the model, as it should be. In practice, his difference of three arcminutes was immaterial.

Godfray's *Lunar Theory* of 1871 discussed the interlinkage between these two functions (Godfray 1871, p. 70; recommended by Whiteside 1976, p. 328, note 47). His equations given for the two above-defined functions are as follows:

$$\delta = 15m/8 \sin 2(\alpha' - \otimes)$$

and

$$E = e[1 + 15m/8 \cos 2(\alpha' - \otimes)],$$

where δ is the second equation of apogee, E is the varying eccentricity, e is mean eccentricity, m is the ratio of lunar tropical month/solar tropical year (0.075), and $(\alpha' - \otimes)$ is the Horrox angle between apse and Sun. This gives maximal values for δ of 8°, and an eccentricity fluctuation of a mere 14%, whereas it has to vary by 21% according to *TMM*. The values are both in deficit by one-third, which is hardly convincing even as a first approxima-tion. Gaythorpe applied the same coefficient $15m/8$ in both of these functions.

A more reliable maximal value for δ was derived by Gaythorpe (1925, p. 859; 1957, p. 136) as $\arcsin(\varepsilon/2e)$, where ε is the coefficient of the evection term (1.274, see below). This is equal to 11°39', using the *TMM* value of eccentricity. The "correct" value of this vitally important constant thereby appears as closer to Horrocks's final value, than to the considerably higher value which Newton, following Halley's advice, gave to it. Gaythorpe derived this value by showing how the modern evection and equation of centre terms were equivalent to a single equation of centre term using an oscillating apse line, δ being its maximal oscillation.

VI. Not the Evection

By thus coupling the libratory motion of the apse line AP with a variable eccentricity, Horrox (and subsequently Flamsteed) united the two principal lunar inequalities: namely, the equation of the centre and the evection. (Forbes 1975, Ch.4, p. 65.)

The first three of the modern equations of the lunar orbit are,

$$6.288 \sin M' \qquad \text{or in our symbols: } \sin (M-A)$$
$$+1.274 \sin (2D-M') \qquad \sin [2(M-S) - (M-A)]$$
$$+0.658 \sin 2D \qquad \sin 2(M-S)$$

where the first term is the elliptic inequality, the second is the evection and the third is the variation. M' represents the Moon's mean anomaly, i.e., distance from its mean apse in longitude, D its mean elongation, i.e., distance from the mean Sun, and M, S and A the mean positions of Moon, Sun and apogee. The Horrocks model conflates the first two of the above equations. How its performance compares with these, is something we will shortly ascertain.

The evection has the characteristic that one cannot picture it, as varying with the sine of twice the elongation minus the anomaly. In this it contrasts with the kinematic model we are considering, which is wholly visual. The evection was named by Ismael Boulliau in 1645; however, its meaning varied somewhat (*GHA*, p. 195). As the "second inequality," it was discovered by Ptolemy, who fixed its maximum value at 1° 19′, described by Dreyer as "very near the true value" (Dreyer 1953, p. 195; Gingerich 1993, p. 59). Its modern meaning, as having a period of $31\frac{1}{2}$ days, developed in the later eighteenth-century.

Chapter 8

The Seven Moons of TMM

>...and the Moon's Place will be equated a seventh time, and this
>is her Place in her proper Orbit. (*TMM*, in Cohen 1975, p. 113)

Newton originated the concept of seven steps of equation as his distinctive approach to lunar theory, in his *TMM* of 1702. He had discerned just seven colours in a rainbow in 1675, and his *Optics* of 1704 found seven steps of colouration in his "Newton's Rings" (P. Gouk in *Let Newton be!* 1988; for the sevenfold pattern of "Newton's rings" see Castillejo 1986, p. 97). Castillejo also noted that *Optics* was composed in seven sections. In like manner Newton found seven steps appropriate for his lunar endeavours. Earlier, Crabtree had described the Horrocksian theory in terms of seven steps, in 1642, as may have influenced him.

This sevenfold structure became a distinctive hallmark of the various "Newtonian" ephemerides that utilised *TMM*. To quote Dr Waff,

>Nearly all *new* lunar tables constructed during the first half of
>the eighteenth century utilised in some fashion his [Newton's]
>tabular theory. (in Cohen 1975, p. 79)

That bold claim by Dr Waff was regrettably never substantiated. It will here be investigated in due course. By "tabular theory" Dr Waff was referring to the Newtonian seven steps of computation. Leadbetter's *Uranoscopia* of 1735 contained the seven steps, Le Monnier's *Institutions* of 1746 in Paris contained them, as (mainly) did Halley's *Astronomical Tables* of 1752 (see Ch. 14 below) Thus, its shadow stretched over half a century, greatly ignored by science historians.

The first summary of *TMM*'s seven steps in trigonometric form was given by Francis Baily, while President of the newly-formed Royal Astronomical Society (Baily 1837, p. 742), as follows:

Table 8.1: Baily's account of the "Newtonian Rules":

I)	11′ 49″	the annual equation
II)	3′ 45″ sin $2(D-A)$	
III)	47″ sin $2(\odot -\Omega)$	
IV)	Equation of centre, including evection	
V)	35′ 15″ sin $2D$	the Variation
VI)	2′ 10″ sin $(2D + a - A)$	
VII)	2′ 20″ sin D	

Baily gave no details beyond this bare outline. He pointed out that four of the equations were entirely new, namely numbers two, three, six and seven. The magnitudes of the sine functions in Baily's summary were mostly mean values, whereas Newton made them vary in relation to several different cycles. (His symbols are different from those that we shall employ: he took D as the Sun-moon angle, Ω as the lunar node, $\odot$ as Sun, and a, A as solar and lunar anomalies).

In the program accompanying this book (available for free download at www.ucl.ac.uk:80/sts/kollrstm/newton.htm) the instructions of *TMM* are translated into a sequence of machine-readable functions. As input, this program takes the time in days after noon GMT on December 31, 1680 (Old Style), and as output it gives lunar longitude. Its latitude function will be described later (Ch. 9, V). Figure 8.1 is a flow-diagram of *TMM*'s sequence of operations. I found that the program based on *TMM* did well accord with the heavens, at least around the time of its composition.

The seven "steps of equation" are here presented as a sequence of additive functions, and are given without explanation. We start with the complete sequence, to show its structure, and then in the next chapter will seek to justify each step. The program starts with a time-value, defining five different mean motions, and these become in turn modified by successive "equations." A sequence of interactions takes place, ending with the moon being equated for a seventh time. Later in the chapter some equivalent algebraic terms are given, for each step.

The sequence here presented contains no twentieth-century astronomical constants, but uses only those given in *TMM*; and, with only one exception, it contains no modern equation: it does include the "equation of centre" formula as discussed in Chapter Six, since *TMM* merely states that tables for the equation of centre were to be compiled, implying that a standard procedure was to be followed, and merely gives maximum and minimum values for it. *TMM*'s instructions on how to accomplish the "reduction," ie conversion to the plane of the ecliptic, are also rather brief, this being a quite standard operation. Thus, with only one exception, what is here presented is merely:

> a translation from the hieroglyphics of geometry into what is
> now the vernacular language of science [i.e., algebra],

—as was claimed by Stevenson's 1834 opus, *Newton's Lunar Theory Exhibited Analytically* (Stevenson 1834, Preface). However, as was indicated earlier (Ch.1, VII), what Mr Stevenson presented was not in fact the Newtonian procedure, but an idealised version thereof, resembling the mid-eighteenth century French theories and quite lacking the Horrocksian mechanism (Cohen 1975, p. 79). Such a translation of *TMM* into "the vernacular language of science" is here accomplished for the very first time.

Checks that were used to test the program have been included as shown below, together with a complete worked example in the form of the case-study by Richard Dunthorne, a Cambridge student who prepared tables which adhered closely to *TMM*. He published this in 1739 as *Practical Astronomy of the Moon: or, New Tables of the Moon's Motions*, the purpose of which was to see how well *TMM* actually worked. Dunthorne put his maximum equation of apogee at 12° 18' 15", as given in the third edition of *PNPM*, whereas *TMM* had given it as 12° 15' 4", that being the sole difference.

While the equations below were all bar one derived from the instructions of *TMM*, I was at times uncertain about the signs, especially for the nodal equations. The worked example given in Dunthorne was invaluable for checking that the addition and subtraction of the trigonometric functions was proceeding correctly. Dunthorne's 1739 opus appears to me as the one work which has embodied 100% the *TMM* rules in its lists of tables and instructions on how to use them. A convention has here been adopted that the faster moving position was subtracted from the slower, for example the solar "anomaly" is represented by $(H - S)$; bearing in mind that $\sin (A - B) = -\sin (B - A)$ and $\cos (A - B) = \cos (B - A)$. The discussion of the four new Newtonian equations given in *GHA* (p. 267) was also of assistance in rightly applying their signs.

The amplitude of the "equation of the centre" is far larger than any other of *TMM*'s "equations." Newton positioned it in the centre of the seven steps, so that there are three antecedent stages and three following. The fifth stage comprises the well-known inequality discovered by Tycho Brahe called "Variation."

I. *TMM* in Machine-Readable Form

The five variables, measured in degrees of zodiac longitude from zero Aries, are: Moon M, Sun S, apogee A, aphelion H and node N. These have motion in degrees/day, and values from zero to 360°, and depend on time t, measured in days from noon G.M.T. Dec. 31, 1680 Old Style. The five variables have starting positions at time zero and speeds as follows:

$$M = 181.763 + 13.17639535 \times t$$
$$S = 290.580 + 0.98564697 \times t$$

$$A = 244.468 + 0.1114083 \times t$$
$$H = 97.392 + 0.0000479 \times t$$
$$N = 174.243 - 0.0529551 \times t$$

These are the mean motions. They are linear functions. For example, putting t equal to 7305, the number of days in twenty Julian years, generates the following positions for TMM's second epoch date (Ch.4, IV).

for $t = 7305$, $M = 315.331$
$$S = 290.731$$
$$A = 338.306$$
$$H = 97.742$$
$$N = 147.406$$

These five longitude values are those specified by TMM for noon on December 31, 1700, confirming that the mean motions tie up with those specified. A "modulo" function retains the value of each function within 0–360°.

The following flow-chart outlines the sequence of interaction of these five variables through the seven steps:

Figure 8.1   The Steps of Equation in TMM

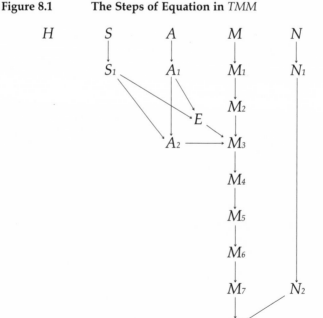

Four Functions:

TMM-PC utilises four functions, whose operation may be outlined as follows:

The eccentricity	f:	$A_1, S_1 \rightarrow E$
Second apse equation	g:	$A_1 \rightarrow A_2$
Equation of centre	h:	$E, A_2, M_3 \rightarrow M_4$
Second node equation	j:	$N_1 \rightarrow N_2$

Function h applies the equation of centre, which gives radian measure (Ch. 6, II) and so required a $180/\pi$ conversion factor to bring it into degrees. The next chapter will explain the derivation of these functions which are as follows:

Eccentricity from Horrox angle:
$$E = f(A - S) \Rightarrow 0.05505 \times \sqrt{1.0454 + 0.4262 \cos 2(A - S)}.$$

Second Apse Equation from the Horrox angle:
$$g(A - S) \Rightarrow \arcsin \frac{\sin 2(A - S)}{85.24 \times f(A - S)}.$$

Equation of Centre from lunar anomaly and eccentricity:
$$h(A - M,E) \Rightarrow [2E \times \sin(A-M) - 1.25 \times E^2 \times \sin 2(A - M)] \times 180/\pi.$$

The *Node Equation*:
$$j(N - S) \Rightarrow \arctan \frac{\sin 2(N - S)}{38.33 + \cos 2(N - S)}.$$

The following test checks whether the functions are working.

For	f,	put	$A - S = 48°$	$\Rightarrow 0.05507,$
	g,	put	$A - S = 48°$	$\Rightarrow 12.23$
	h,	put	$A - M = 30°$ and $E = 0.05$	$\Rightarrow 2.717$
	j,	put	$N - S = 120°$	$\Rightarrow -1.311$

The Seven Steps

The steps of equation are inserted in accord with the above flow-diagram. Thus, the apogee first-equated (A_1) feeds into functions f and g, then adding the value of g to A_1 gives the apogee second-equated. This in turn feeds into function h, the equation of centre, to give M_4. The node only receives its second equation at a late stage, for performing the reduction. The Horrox-angle between Sun and apse (Ch. 2, III) we now define as being between the first-equated positions, i.e., $(A_1 - S_1)$.

Step One: the annual equation.

This step begins from the "mean motions," linearly time-dependent functions with modulo 360°, as given earlier.

$$S_1 = S + 1.939 \times \sin(H - S) - 0.0205 \times \sin 2(H - S)$$
$$M_1 = M - 0.197 \times \sin(H - S)$$
$$A_1 = A + 0.333 \times \sin(H - S)$$
$$N_1 = N - 0.158 \times \sin(H - S).$$

Step Two: $M_2 = M_1 + \left[6.25 - 0.31 \times \cos(H - S_1)\right] \times \sin 2(A_1 - S_1) \sqrt{100}.$

Step Three: $M_3 = M_2 + 0.0131 \times \sin 2(N_1 - S_1).$

Step Four: Put $E = f(S_1, A_1)$
          and $A_2 = A_1 - g(S_1, A_1);$
          then $M_4 = M_3 + h(E, M_3, A_2).$

Step Five: the Variation.

$$M_5 = M_4 + \left[0.5923 - 0.0312 \times \cos(H - S_1)\right] \times \sin 2(M_4 - S_1).$$

Step Six: $M_6 = M_5 + 0.0361 \sin(S_1 - M_5 + H - A_2).$

Step Seven: $M_7 = M_6 + \left[0.0389 + 0.015 \times \cos(H - A_2)\right] \sin(S_1 - M_6).$

Reduction: Put $N_2 = N_1 - j(S_1, N_1);$
      then $M(end) = M_7 + 0.1160 \times \sin 2(N_2 - M_6)[1 + 0.0586 \cos 2(N_2 - S_1)].$

The following test checks the entire sequence of equations, utilising Dunthorne's worked example (Dunthorne 1739, pp. 50–59; Table 8.2), which took the instant of 3.40 pm on January 2nd 1737. This position has been used as a standard test for setting up the program. The following positions were generated by TMM-PC for this instant:

for t = 20456.1528 days,

M =	80.119		M_1 = 80.069
S =	293.124		M_2 = 80.112
A =	3.453		M_3 = 80.124
H =	98.372	E = 0.04670	M_4 = 74.815
N =	170.985		M_5 = 74.207
			M_6 = 74.186
S_1 =	293.628		M_7 = 74.163
A_1 =	3.538		M_{end} = 74.142 = 74° 8′ 31″
A_2 =	354.212		
N_1 =	170.945		
N_2 =	169.572		

One can be a day out in computing the t-value owing to leap-years, and so the solar values are first checked to see if they concur. The spreadsheet has the form of a flow diagram, where from the t-value inserted at the top, all the other values are defined. It shows merely the figures generated at each stage, but not the functions that produced them.

The mean motions concur with those of Dunthorne within an arcsecond, confirming what was said in Chapter Five that "the Newtonians" in this period based their mean motion tables firmly upon *TMM*. The programme requires *TMM*'s value for lunar tropical motion to be defined to ten figures as

$$13.17639535°/\text{day}.$$

That level of accuracy is vital if results are to be quoted to arcseconds. It differs from the then "true" value, that is to say as interpolated into historical time using Meeus's modern values (Ch. 5, II), of

$$13.1763967°/\text{day}.$$

The difference appears in the sixth place of decimals. For the slower, solar mean motion, eight figures were sufficient for arcsecond accuracy. For comparison, the equations for locating correct lunar mean motions in historical time (Appendix II) require an eleven-figure term.

The final results differ in ecliptic longitude by eighteen arcseconds, which is tolerable. Richard Dunthorne was an eminent astronomer in his own right: it was he who first established Edmond Halley's conjecture of the secular acceleration of the Moon. Halley had proposed in 1696 that such an effect was causing eclipses in antiquity to be displaced by an hour or so from their expected times, but he never showed any computations on the matter. Dunthorne did this, and Brewster's *Memoirs* of Isaac Newton referred to Dunthorne in this context. (Brewster 1855, Vol. I p. 355)

We can inspect the "equations" for each of the seven stages, comparing Dunthorne's with those of TMM-PC (see over). The largest discrepancy is ten arcseconds, in the sixth equation. A "correct" answer is given from a modern program, showing how final values err by nearly seven minutes of arc, which is rather shocking considering that it is thrice the maximum error claimed by Gregory for *TMM* in 1702. Later on, we will see how often *TMM* would generate an error of such magnitude.

	Dunthorne	*TMM-PC*			*Dunthorne*	*TMM-PC*
Eqn (1)	$-3'\,2''$	$-3'\,0''$	1st node eqn:		$-2'\,27''$	$-2'\,24''$
(2)	$+2'\,33''$	$+2'\,32''$	2nd node eqn:		$-1°22'\,23''$	$-1°22'\,23''$
(3)	$+43''$	$+42''$				
(4)	$-5°\,18'\,29''$	$-5°\,18'\,30''$	1st apse eqn:		$5'\,9''$	$5'\,5''$
(5)	$-36'\,28''$	$-36'\,28''$	2nd apse eqn:		$9°\,19'\,27''$	$9°\,19'\,34''$
(6)	$-1'\,27''$	$-1'\,17''$				
(7)	$-1'\,28''$	$-1'\,20''$	eccy. ($\times 10^6$):		46703	46705
Reduction		$-1'\,17''$	$-1'\,16''$			
Final ans.	$74°\,8'\,13''$	$74°\,8'\,27''$				
Correct value:		$74°\,15'\,3''$				

The above "correct" value was obtained using a copy of the I.L.E. program kindly supplied by Dr. Bernard Yallop at the Royal Greenwich Observatory, said to be accurate to within a second or two of arc in historical time, which contains sixteen hundred terms.

Omitting the mean motions, we can cast the central chain of equations into a more algebraic format, as follows (the essential parameters are listed at end of chapter). Concerning the signs of the functions, suppose for example one were doubtful about that present in "$1 - 3E \cos (H - S)$", a term which appears in the second and fifth stages and represents an annual fluctuation about a mean value. *TMM* states that this "equation" has to be maximal at perihelion (near to midwinter) and minimum at aphelion. To check the correctness of the expression, a date when the Sun is near aphelion was inserted, giving S and H similar values, when the expression should reach its minimum value, while conversely it should rise to a maximum value when 180° or thereabouts separates S and H. This was found to be the case.

II. The Seven Steps as Trigonometric Functions

Step One: the solar and lunar equations of centre:

$$S_1 = S + 180/\pi[2e \sin (H - S) - 1.25e^2 \sin 2(H - S)]$$

where e is solar eccentricity, 0.01692;
180/π converts radians to degrees;

$$M_1 = M - 11' \, 49'' \sin (H - S)$$

where M is mean lunar longitude.

Step Two: $M_2 = M_1 + 3' \, 45''[1 - 3e \cos (H - S_1)] \sin 2\Phi$

Step Three: $M_3 = M_2 + 47'' \sin 2(N - S_1)$

Step Four: the lunar equation of centre:

In the figure, TC is unity, and radius CF has length $\varepsilon = 0.2131$, as half the difference between maximum and minimum eccentricities divided by the mean value. Then TF represents the varying eccentricity E, FTC is δ, the equation of apogee, and FCB is 2Φ, twice the Sun-apse angle. Then,

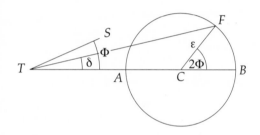

$$E = 0.05505\sqrt{(1 + \varepsilon^2 + 2\,\varepsilon \cos 2\Phi)},$$

cosine formula on ΔFTC; and

$$\sin \delta = \sin 2\Phi \times \frac{0.01173}{E},$$

sine formula on ΔFTC.

(0.011732 is half the difference between maximum and minimum eccentricity, 0.055050 is the mean.)

$$A_2 = A_1 - \delta,$$

and

$$M_4 = M_3 + 180/\pi[2E \sin (A_2 - M_3) - 5/4E^2 \sin 2(A_2 - M_3)].$$

Step Five: the Variation

$$M_5 = M_4 + 35' \, 32''[1 - 3e \cos (H - S_1)] \sin 2(M_4 - S_1).$$

Step Six: $M_6 = M_5 + 2' \, 10'' \sin (S_1 - M_5 + H - A_2).$

Step Seven: $M_7 = M_6 + [2' \, 20'' + 54'' \cos (H - A_2)] \sin (S_1 - M_6).$

The sign of the seventh equation is confusing, since one version of *TMM* (Cohen 1975a, p. 113) specified that it be additive for the waxing Moon and subtractive for the waning, while another version a few pages later (Cohen 1975a, pp. 138–9) specifies the converse. The latter version has been used by *GHA*, and is indeed the correct way round in accord with modern equations. Both these versions were published in 1702, but no one, not even Flamsteed or Baily, has remarked upon this divergence. The sign of the sixth equation was later reversed, as Flamsteed pointed out (Cohen 1975a, p. 59), but not the seventh. Its sign as above is negative for the waxing Moon.

III. A Comparison with Flamsteed

Table 8.2 shows Dunthorne's mode of summarising his computation. By contrast, the customary format for these matters, prior to *TMM*, is shown by a computation example as given in *DOS* (p. 38), Table 8.3. More than half a century separates these two case-studies; indeed, the arrival of the *Principia* separates them. How do they compare?

DOS presented ten steps for finding lunar longitude. It began with the equation of time, converting solar into mean time—a stage strangely omitted by *TMM*. The proud claim was made that:

> For he [the author, Flamsteed] will not dissemble it, that tho he esteems these [principles] far better than any yet published; he is sensible that the solar may be some little faulty, but scarce more than a Minute; the lunar he finds often to Err 5 or 6 Minutes, and sometimes (tho rarely, and at most) 10 or 11 minutes; which yet he can the easier bear, whilst he sees the Numbers of other more famous and celebrated Men to err 15 or 16 minutes, at the same time when his agree nearly with the Heavens. (*DOS*, p. 34)

As ill-luck would have it, this declared maximum possible error turned up in the sole example to illustrate Flamsteed's theory! Its true place of the

Table 8.2: Lunar longitude computation by Dunthorne (1739) using a procedure identical to that of *TMM*. For 3.40 pm on Jan. 2nd, 1737, he found 74° 8′ 13″, the correct value being 74° 14′ 29″.

	Equ. Time	Longitude ☽	Apogee ☽	Asc. Node ☽
		s o ′ ″	s o ′ ″	s ° ′ ″
1721	1721	2 28 53 55	2 12 8 35	4 0 34 5
16	16	10 22 51 16	9 21 4 12	10 9 28 12
January		0 0 0 0	0 0 0 0	0 0 0
d.2		0 26 21 10	0 13 22	0 6 21
h.3		1 38 49	0 50	0 24
′40		21 58	11	5
mean Mot		2 20 7 8	0 3 27 10	10 9 35 2 · 5 20 59 3 ♌
1 *Equation*		− 3 2	+ 5 9	− 2 27
☽ 1 *Equa.* / 2 *Equation*		2 20 4 6 / + 2 33	0 3 32 19 / 9 23 37 22	5 20 56 36 ☊ 1 *Eq.* · 9 23 37 22 *Suns Pl.*
☽ 2 *Equa.* / 3 *Equation*		2 20 6 39 / + 43	9 20 5 3 / − 9 19 27	4 2 40 46 *An. Arg.* · − 1 22 2 *Eq. ☽*
☽ 3 *Equa.* / 4 *Equation*		2 20 7 22 / − 5 18 29	11 24 12 52 / 2 25 54 30 *mean Anom.*	5 19 34 13 ☊ 2 *Eq.* [21° 3′]
☽ 4 *Equa.* / 5 *Equation*		2 14 48 53 / − 36 28	*Excentricity* 46703	8 24 35 17 *Arg. Lat.* · 5 4 40 *Inc. Lim.*
☽ 5 *Equa.* / 6 *Equation*		2 14 12 25 / − 1 27	4 21 11 31 / 9 12 23 2	☽ *à ☉* · *doubled*
☽ 6 *Equa.* / 7 *Equation*		2 14 10 58 / − 1 28	3 8 22 18 / 8 15 50 34	*Apogee ☉* · *Ap. ☽ à Ap. ☉*
☽ *in Orbit* / *Reduction*		2 14 9 30 / − 1 17	1 7 2 5	*Arg.* 6 *Equation*
☽ *in Eclip*		2 14 8 13	←	
Lat. South		5 3 18		

Table 8.3: a similar computation by Flamsteed (*DOS*, p. 38) for 6.35 pm, Dec. 22nd, 1680. He obtained 5° 9′ 52″ of Gemini, the correct answer being 4° 59′.

	Mean Motion.	Apogee.	Node Retrog	
	s o ′ ″	s o ′ ″	s o ′ ″	
1661	01 18 10 14	5 00 21 51	6 21 04 47	Radical place of the Node.
19	11 21 00 07	1 23 03 29	0 07 27 20	
December 22B	23 58 22	1 09 46 22	18 54 19	
h. 6	3 17 39	1 40	48	
′ 34	18 40	10	5	
″ 57	31	8 03 13 32	0 26 22 32	Motion of the ♌ from the Radix.
Mean Motion	02 06 45 33	9 12 09 35	5 24 42 15	♌ true place.
Physical parts sub.	1 03	1 08 56 03	AnnualArgument.	
Mean Motion Cor. Apoge	02 06 44 30	10 50 32	Equat. of the Apoge add.	
	08 14 04 04	8 14 04 04	True place of the Apoge.	
Mean Anomaly / Equation subtract.	05 22 40 26 / 54 22	Ex. 57678 / M.Ex. 55237	Eq. under the	Greatest Excen. 01 03 52 } Diff. 12′ 1″. / Mean Excenter 00 51 51 }
☽ Equation place / Suns	02 05 50 08 / 09 12 09 35	Diff. 2441		Part propo. add 00 02 31 / Absolute Equat. 00 54 22
☽ from the Sun / Variation subtract	04 23 40 33 / 36 16	♌ Mean pla. / Suns	5 24 42 15 / 9 12 09 35	R.c.s. 5 01 36 9,998327 / t. 71 29 22 10,475214
☽ in her Orb	02 05 13 52	☊ from the ♌	3 17 27 20	t. 71 25 22 10,473541
Node	05 23 44 30	Equat. ♌ sub.	57 45	Red. sub. 4 00
Argument of Lat. / Reduction Subtra.	08 11 29 22 / 4 00	♌ true place / In. of the Orb	5 23 44 30 / 5 01 36	R.s. 5 01 36 8,942600 / s. 71 29 22 9,976930
☽ place in the Eclip / Latitude South	Ⅱ 05 09 52 / 04 45 58	←	s. 4 45 58 8.919530	
Moons Horiz. Paral / Horizontal Semid.	61 31 / 16 4		Hence the M. true place is Ⅱ 5 09 52 / Latitude South 4 45 58	

Moon was eleven minutes in advance of what it should have been: at 6.35 pm GMT, on December 22 1680 Old Style, its longitude was 4° 59′ of Gemini, compared with the computed value of 5° 10′ found in *DOS*. *DOS*'s mean Moon positions were two minutes *behind* their proper values (Chapter 5), so it would appear as if his ten steps had introduced no less than thirteen minutes of error.

The accuracy of this example accords with the pessimistic view that Flamsteed expressed in the *Philosophical Transactions* of 1683, after he had been the Astronomer Royal for seven years:

> The best tables of the Moon's Motions do err 12 minutes or
> more, in her Apparent Place. (*Phil. Trans.*, Vol.13, p. 405)

In his view, methods based upon the moons of Jupiter offered the best means of finding longitude, as using the lunar method "the calculations will be so perplexed and tedious." This view expressed by Britain's Astronomer Royal was quoted in John Harris's *Lexicon Technicum* of 1704, so may well have expressed a general view. The research of Owen Gingerich (Ch. 1, pp. 19–20) entirely confirmed this assessment, finding indeed that larger errors were common in ephemerides of the period.

Let us compare the accuracy of these two worked examples half a century apart, for the three variables of solar, lunar and node positions.

Source	*Moon position*	*Sun position*	*node position*
1) *DOS*	5° 09′ 52″ Gemini	12° 09′ 35″ Capr.	23° 44′ 30″ Virgo
True posns:	4° 59′ 18″	12° 08′ 00″	24° 01′ 16″
Errors:	+10′ 32″	+01′ 35″	−16′ 45″
2)Dunthorne	14° 8′ 13″ Gemini	23° 37′ 22″ Capr.	19° 34′ 13″ Virgo
True posns:	14° 15′ 00″	23° 37′ 27″	19° 49′
Errors:	−6′ 47″	−05″	−14′

1) 22 December, 1680 at 6.35pm GMT, London.
2) 2 January, 1737, 3.40 p.m., "Time equated."

Errors are measured here and throughout as {historic-true} values. Only for solar longitude is the position to seconds of arc at all relevant.

Dunthorne's "to the Reader" does not extol *TMM*'s accuracy, but rather admits "that the Newtonian Numbers are a little deficient." The above figures suggest a mild improvement over half a century. The *DOS* solar error of one and a half arcminutes is surprisingly large, considering that, as we saw in Chapter

Five, his mean motion was within arcseconds at this period. Flamsteed's solar numbers were improved several times after *DOS*'s composition.

A summary of the constant terms given for the equations of *TMM* appears in Table 8.4.

IV. A Test of Accuracy

The accuracy of *TMM* was investigated using the above computer program (hereinafter referred to as TMM-PC), by comparing its results against a modern ephemeris program accurate to seconds of arc. Noon values of longitude on successive days of December 31st (Old Style) were taken, sampling at two year intervals over a period of six decades, 1680–1740. TMM-PC measures time from noon on December 31st 1680, which means that the initial reading was at time zero, then the next was for 730.5 days, and so forth. Both solar and lunar longitude values were read off from the

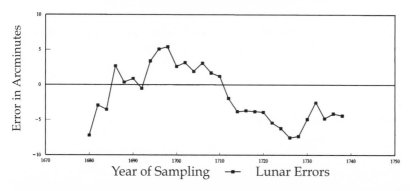

Accuracy Test of *TMM*
Sampling Two-Yearly

Year of Sampling   —■—   Lunar Errors

Figure 8.2: Sampling from Dec.31st noon GMT, Old Style, two-yearly (every 730.5 days) showing errors in arcminutes over six decades.

program, the former being necessary to check that the number of days inserted was correct, since an extra day from a leap year shows up as a degree displacement in longitude. The program to obtain the longitudes was checked against standard values obtained from the R.G.O. The results obtained over six decades are depicted graphically in Figure 8.2.

There is a slight drift in the baseline of *TMM* through the decades, as cumulative error of its mean motion. The overall error-values in arcminutes were -1.6 ± 3.8 for the moon and 0.2 ± 0.3 for the Sun. A long-term pattern appears as present in the data, of period fifty years or so, which is a consequence of the sampling interval used being a multiple of a major *TMM* period, viz the year. Chapter eleven will treat this issue more thoroughly.

V. Halley's Judgement

Halley's mature and final opinion on the subject was given in 1731, when he was Britain's foremost astronomer and both Newton and Flamsteed were mere memories for him. Then, after consulting both his own lunar tables (as Astronomer Royal) and those of his predecessor, his view of *TMM* three decades after its composition was that:

> The Faults of the Computus formed therefrom rarely exceed a quarter Part of what is found in the best Lunar Tables before that time extant. ...
>
> By this it was evident that Sir Isaac had spared no Part of that Sagacity and Industry peculiar to himself, in settling the Epoches, and other Elements of the Lunar Astronomy, the Result many times, for whole Months together, rarely differing two Minutes of Motion from the Observations themselves.
> (*Phil. Trans.* 1731/2, p. 191)

This comment was made a propos of the 1713 version given in *PNPM* (pp. 655–664), which Halley viewed as an improvement upon *TMM*. These remarks of his echo what he had written years earlier in 1716, in a Foreword to Streete's *Astronomia Carolina* that he re-issued (Streete 1716). Considering the above-discussed verdict of Britain's first Astronomer Royal, that even the best lunar ephemerides were liable to err by twelve minutes "or more," it seems likely that *TMM* was capable of delivering a slight enhancement of predictive power, though it hardly achieved that which Halley claimed for it. Dunthorne was not over-impressed by its accuracy (Dunthorne 1739, Ch. 8 Section 3).

Later, we will study the various Newtonian ephemerides which modelled themselves upon *TMM*, and ascertain whether they achieved a superior predictive accuracy to others such as those of Jacques Cassini in Paris, who did not use it.

Table 8.4:   The Constants of *TMM*

(i.e., the maximal or peak amplitude-values)

The annual equation (I)	
Sun	1° 56′ 20″
Moon	11′ 49″
apogee	20′ 0″
node	−9′ 30″
Equation (II)	
maximum (in winter)	3′ 56″
minimum (in summer)	3′ 34″
Equation (III)	47″
The equations of the center	
Sun	1° 56′ 20″
Moon	7° 39′ 30″
Equation of apogee (IV)	12° 15′ 4″
Eccentricity	
maximum (apse conjunct Sun)	0.066782
minimum	0.043319
The Variation (V)	
maximum (in winter)	37′ 25″
minimum (in summer)	33′ 40″
Equation (VI)	2′ 10″
Equation (VII)	2′ 20″
Angle to ecliptic	
maximum (nodes conjunct Sun)	5° 17′ 20″
minimum (quadrature)	4° 59′ 35″

Chapter 9

Commentary on TMM, *Continued*

This chapter continues the Commentary from Chapter 4, and indicates how the equations just described were obtained from *TMM*. This involves summarising the content of previous chapters, while avoiding repetition wherever possible.

I. An English Pamphlet

The quotations from *TMM*, here as in previous chapters, come from the first English edition, published in 1702. This edition had no editor or translator specified, but a Preface attached to it contained remarks indicating that it appeared shortly after Gregory's *Astronomiae Elementa*, containing a Latin version of *TMM*, which appeared in that same year. Bernard Cohen found no definite evidence from any correspondence or contemporary comment that Halley was the author of this Preface (Cohen 1975, p. 32)— an idea originally proposed by Augustus De Morgan in the nineteenth-century (Cohen 1975, p. 30); yet, it remains plausible from considerations of Halley's style, and familiar manner towards the persons concerned, viz. Newton and Gregory. Its brief Preface praised Gregory's book and added that, since many would not be able to afford it, the pamphlet would be convenient. It is not necessarily a translation from Gregory's Latin, since, as Cohen argued, the original version of *TMM* was probably in English. The manuscript is identical in content with that published by Gregory, except for a divergence in the seventh equation discussed below, unnoticed by Cohen.

II. The Annual Equations

TMM conferred annual inequalities upon four of its ecliptic variables, the node, perigee, Sun and Moon, the first two being innovative:

> These mean Motions of the Luminaries are affected with various inequalities: Of which,
>
> 1. There are the Annual Equations of the aforesaid mean Motions of the Sun and Moon, and of the Apogee and Node of the Moon.
>
> The annual Equation of the mean Motion of the Sun

depends on the Eccentricity of the Earth's Orbit round the Sun, which is $16\frac{11}{12}$ of such parts, as that the Earth's mean Distance from the Sun shall be 1000: Whence 'tis called the *Equation of the Centre*; and is when greatest 1°.56'.20".

The greatest Annual Equation of the Moon's mean Motion is 11'.49". of her Apogee 20'. and of her Node 9'.30".

The Equation of Center derived from Flamsteed's value of 1692. There is an exact equivalence between it and *TMM*'s eccentricity values, though the method by which this computation was then performed is unclear. Chapter Six looked at how the modern equation of center links the two together. In the case of the Earth's orbit, the function reaches a maximum value at an anomaly *M* of 91°. Inserting it in the equation together with an eccentricity value of $16\frac{11}{12}$ parts in 1000 (i.e., 0.016917) gave an agreement within one second of arc! This Equation differs from what was then the true value by 45 arcseconds, while the value given for eccentricity differs from the modern value by 0.5% (Appendix I).

TMM comments further about the interlinking of these annual equations, of interest as showing how tricky such things were before trigonometrical formulae became available:

And these four Annual Equations are always mutually proportional one to another: Wherefore when any of them is at the greatest, the other three will be greatest; and when any one lessens, the other three will also be diminished in the same *Ratio*.

The Annual Equation of the Sun's Centre being given, the three other corresponding Annual Equations will be also given; and therefore a Table of *That* will serve for all. For if the Annual Equation of the Sun's Centre be taken from thence, for any time, and be called *P*, and let *P*/10 = *Q*, *Q* + *Q*/60 = *R*, *P*/6 = *D*, *D* + *D*/30 = *E*, and *D* − *D*/60 = 2*F*; then shall the Annual Equation of the Moon's mean Motion for that time be *R*, that of the apogee of the moon will be *E*, and that of the Node *F*.

Only observe here, that if the Equation of the Sun's Centre be required to be *added*; then the Equation of the Moon's mean Motion must be *subtracted*, that of her apogee must be *added*, and that of the node *subtracted*. And on the contrary,

> if the Equation of the Sun's Centre were to be *subducted*, the
> Moon's Equation must be *added*, the Equation of her Apogee
> *subducted*, and that of her node *added*.

These four functions vary as the sine of solar anomaly, so are maximal near the equinoxes and zero near the solstices. The four equations we extracted from these instructions were:

$$S_1 = S + 1.939 \sin (H - S) - 0.0205 \sin 2(H - S)$$
$$M_1 = M - 0.197 \sin (H - S)$$
$$A_1 = A + 0.333 \sin (H - S)$$
$$N_1 = N - 0.158 \sin (H - S)$$

The solar annual equation is the only one large enough to merit a second term of the equation of centre in the *TMM* program. It is evident that these constant terms are linked through the ratios specified by *TMM*, e.g.: $R = 61/60 \times P/10$ (since $Q = P/10$ and $R = 61Q/60$) where P is the maximal solar equation of 1.939, giving $R = 0.197$. These ratios nowadays appear as superfluous, as the amplitudes of the four annual equations have already been given.

Such are the positions "first equated" in *TMM*'s terminology, meaning fluctuations of yearly periods around the mean motions. The Sun has only the one equation, whereas the node and apogee receive second equations at later stages.

III. Two New Equations

The first of *TMM*'s new equations now appears:

> There is also an *Equation of the Moon's mean Motion*
> depending on the Situation of her Apogee in respect of the
> Sun; which is *greatest* when the Moon's apogee is in an
> Octant with the Sun, and is nothing at all when it is in the
> Quadratures or Syzygies. This Equation, when greatest, and
> the Sun *in Perigeo*, is 3′.56″. But if the Sun be in *Apogeo*, it
> will never be above 3′.34″. At other distances of the Sun from
> the Earth, this equation, when greatest, is reciprocally as the
> Cube of such Distance.
>
> But when the Moon's Apogee is any where but in the
> *Octants*, this Equation grows less, and is mostly at the same
> distance between the Earth and the Sun, as the Sine of the
> double Distance of the Moon's Apogee from the next

Quadrature or Syzygy, to the Radius.
This is to be *added* **to the Moon's Motion, while her Apogee**
passes from a Quadrature with the Sun to a Syzygy; but is to
be *subtracted* **from it, while the Apogee moves from the**
Syzygy to the Quadrature.

The function is given as varying with the Horroxian year, which we have designated as the $(A - S)$ function, marking solar conjunctions with the apse, of period 411 days. The function evidently varies as $\sin 2(A - S)$, peaking at the octants, i.e., at the 45° angles, since it passes through two maxima and minima per revolution. The phrase, "nothing at all when it is the Quadratures or Syzygies" implies a sine function crossing its baseline four times per cycle.

While the annual equations had constant coefficients, here the amplitude itself varies during the course of the year, being maximal at perihelion in midwinter. Multiplying by

$$1 - 0.0489 \cos (H - S)$$

will accomplish this. As a cosine function its maxima and minima fall on the aphelion-perihelion axis, and give the required range of ± 11″ about a mean value of 3′ 45″ as specified. Expressing this mean amplitude as 0.0625°, the overall expression becomes:

$$M_2 = M_1 + 0.0625 \left[1 - 0.0489 \cos (H - S_1)\right] \times \sin 2(A_1 - S_1)$$

Feeding the first-equated value M_1 into this expression will generate *TMM*'s second "Equation" of the Moon. Halley called this second equation *aequatio prima semestris*. Baily's comment was, "We have nothing equal to it in amount (depending on the same argument) in the tables of Mayer, Burgh, or Burckhart" (Baily 1837, p. 742). On the other hand, Curtis Wilson expressed the view that, of the four new equations, it was the only correct one (*GHA*, p. 267). The next chapter will resolve this question.

The phrase "reciprocally as the cube of distance" contains an echo of gravity theory, the sole trace in *TMM*, which was to be much developed in the second edition of *PNPM*. The Sun-Earth distance varies by ±1.70% in the course of a year, so a cubic function of this distance would generate thrice this, which is ± 5.10%. The second and sixth equations have their amplitudes modified to vary inversely as the cube of the Earth's distance from the Sun: for equation 2, the given amplitude fluctuation comes to ± 4.9% of its mean value, while for equation 6 it is ± 5.3%, which is tolerably close to the inverse-cube relationship. Our *TMM* program has used these latter values.

These fluctuations are small changes in a three arcminutes function, so any such differences are immaterial.

Curtis Wilson (*GHA*, p. 267) gave the second equation as:

$$-3' \, 45'' \, [1 - 3E \cos (S - H)] \sin 2(S_1 - A_1)$$

as is fairly similar to that given above. Here $(S - H)$ is the Earth's "true anomaly." Inserting the earth-orbit eccentricity into Wilson's term gives an amplitude modification of 5.1% as his "$3E$" term, for both the second and sixth equations. Wilson derived his term from the instruction, "inversely as the cube of the difference," while we have simply taken the amplitude variations specified, the results being similar. Wilson in *GHA* (p. 267) also assigned an inverse-cube amplitude modulation to the third equation, in which we do not follow him: it is merely the second and sixth equations which have this adjustment.

The third equation introduces the nodes:

> **There is moreover another *Equation of the Moon's Motion*, which depends on the Aspect of the Nodes of the Moon's Orbit with the sun: and this is *greatest* when her Nodes are in *Octants* to the Sun, and vanishes quite, when they come to their Quadratures or Syzygies. This Equation is proportional to the sine of the double Distance of the Node from the next Syzygy or Quadrature; and at greatest is but 47". This must be added to the Moon's mean Motion, while the Nodes are passing from their Syzygies with the Sun to their quadratures with him; but subtracted while they pass from the Quadratures to the Syzygies.**

This function passes through two cycles per solar revolution against the nodes. Its amplitude is fixed, and we readily ascertain its formula to be

$$M_3 = M_2 + 0.0131 \sin 2(N_1 - S_1).$$

To ascertain the signs of these functions, we recall that

$$\sin(A - B) = -\sin (B - A),$$

whereas

$$\cos (A - B) = \cos (B - A).$$

Interpretations of whether a sign should be added or subtracted have been checked against the worked example of Dunthorne. The instruction that a function is additive "while the Nodes are passing from their Syzygies with the Sun to their Quadratures with him," and subtractive for the converse, is interpreted as $-\sin 2(S - N)$, or $\sin 2(N - S)$.

IV. The Horrox-Wheel Mechanism

The fourth equation is by far the largest of the seven steps. The deferent-wheel invented by the young Horrocks in 1638 here generates both the eccentricity fluctuation and the apse-line motion, as it revolves once per $6\frac{1}{2}$ months. We have just seen how *TMM*'s value for solar eccentricity tied up exactly with its Equation of Centre value, as implies a definition identical with the modern one of $b^2 = a^2(1 - e^2)$, where a and b are the major and minor axes of an ellipse. The eccentricity e thereby defined is the distance between focus and centre divided by a, the radius of a circumscribing circle (Ch.2, IV).

From the Sun's true Place take the equated mean Motion of the Lunar Apogee, as was above shewed, the Remainder will be the Annual Argument of the said Apogee. From whence the *Eccentricity of the Moon*, and the *second Equation* of her Apogee may be compar'd after the manner following (*which takes place also in the Computations of any other intermediate Equations*).

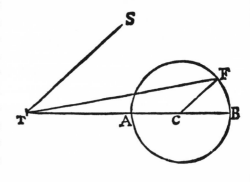

Referring to the diagram, the first sentence defines $(S_1 - A_1)$, represented by the angle *STA*. What is here called the Annual Argument must not be confused with the Annual Equation, discussed earlier. The explanation given in Chapter 7 avoided the term Annual Argument, as liable to confuse, instead calling it the Horrox angle. After all, the cycle is only quasi-annual. I have not grasped the meaning of the final phrase in brackets.

We now come to the instructions for operating the Horrox-wheel diagram, recalling, despite appearances, that the angle *FCB* is supposed to be twice the size of *STA*.

Let *T* represent the Earth, *TS* a Right Line joining the Earth and Sun, *TACB* a Right Line drawn from the Earth to the middle or mean place of the Moon's Apogee, equated, as above: Let the Angle *STA* be the Annual Argument of the aforesaid Apogee, *TA* the least Eccentricity of the Moon's Orbit, *TB* the greatest. Bisect *AB* in *C*; and on the Centre *C* with the Distance *AC* describe a Circle *AFB*, and make the angle *BCF* = to the double of the Annual Argument. Draw the Right Line *TF*, that shall be the Eccentricity of the Moon's Orbit; and the angle *BTF* is the second Equation of the Moon's apogee required.

The revolution of *F* around the circle twice per Horrox-year defines two functions, which are thereby mathematically linked: the second equation of the apse line, *FTA*, with a maximum of twelve degrees, and the eccentricity of the lunar orbit as the length *FT*. The dimensions of the Horrox-wheel are specified as follows:

In order to whose Determination let the mean Distance of the Earth from the Moon, or the Semidiameter of the Moon's Orbit, be 1000000; then shall its *greatest Eccentricity TB* be 66782 such Parts; and the *least TA*, 43319. So that the greatest Equation of the Orbit, viz. when the Apogee is in the Syzygies, will be 7°.39′.30″. or perhaps 7°.40′. (for I suspect there will be some Alteration according to the position of the apogee in Cancer or Capricorn.) But when it is in Quadrature to the sun, the greatest Equation aforesaid will be 4°.57′.56″. and the greatest Equation of the Apogee 12°.15′.4″.

This is innovative, being the first time that these two functions had been so defined, as derived from the same geometry. The modern equation of centre enables us to check what *TMM* calls the Equation of Orbit, which is the amount whereby the Moon's position diverged from the mean motion, maximal at seven and a half degrees.

The magnitude of the apse equation here specified is considerably larger than that specified by DOS, about four percent more in fact. In Newton's letter of April 23rd 1695 a table for what he called the "Annual Argument" gave an eccentricity function virtually unchanged from Flamsteed's, while the apse equation's maximum value has been increased from 11° 47′ to 12° 10′. It is here increased further, and will reach Newton's maximum value of 12° 18′ in the *PNPM* of 1713. Gaythorpe showed how this Newtonian value was more than half a degree larger than was warranted by its modern equivalent, the evection term (Ch. 7, V).

The functions which the previous chapter called *f, g* and *h* accomplish these steps. The first of these obtains the eccentricity, as the length *FT*, given the angle *FCB* as $2(S - A)$ and the lengths *FC* and *TC* as 1173 and 55050 as parts per million, by applying the cosine rule to triangle *FTC*:

$$\text{Eccentricity} = f(A - S) = 0.05505 \times \sqrt{1.0454 + 0.4262 \cos 2(A - S)}$$

As a cosine function it will make eccentricity peak at zero and 180° Horrox angles, i.e., the solar conjunctions to the apse. Then, applying the sine rule to the triangle *FTC*, the angle *FTC* as the second equation of apogee is found by the function $g(A - S)$, whose maximum value is 12° 15′.

The *TMM* text expresses the suspicion, that the apse position in relation to the aphelion line has some effect, but no such long-term cycle was embodied into its procedure.

Having from these Principles made a Table of the Equation of the Moon's Apogee, and of the Eccentricities of her Orbit to each degree of the Annual Argument, from whence the Eccentricity *TF*, and the Angle *BTF* (*viz.* the second and principal Equation of the Apogee) may easily be had for any Time required; let the Equation thus found be added, to the first Equated Place of the Moon's Apogee, if the annual Argument be less than 90°, or greater than 180°, and less than 270°; otherwise it must be subducted from it: and the sum or Difference shall be the Place of the Lunar Apogee secondarily equated; which being taken from the Moon's Place equated a third time, shall leave the mean Anomaly of the Moon corresponding to any given Time. Moreover, from this mean Anomaly of the Moon, and the before-found Eccentricity of her Orbit, may be found (by means of a Table of Equations of the Moon's Centre made to every degree of the mean anomaly, and some Eccentricitys, *viz* 45000. 50000. 55000, 60000 and 65000) the *Prostaphaeresis* or Equation of the Moon's Centre, as in the common way: and this being taken from the former Semicircle of the middle anomaly, and added in the latter to the Moon's Place thus thrice equated, will produce the Place of the moon a fourth time equated.

The lunar Equation of Centre was required for preparing tables, and here the instructions are to prepare them with anomaly angle versus eccentricity values, using five columns of differing eccentricities as compared with DOS's three columns. Some astronomers of the first half of the eighteenth century did follow this advice, e.g. Le Monnier in Paris. Interpolating between, say, anomaly values at one degree intervals and several eccentricity values was not in itself easy. In Chapter Eight, we saw how the main error in Dunthorne's worked example came from this fourth step of equation, creeping in during his interpolations over the Equation of Centre table.

Our function h is the modern formula for equation of centre, using the eccentricity and equated apse position as produced by the Horrox-wheel. Our method uses the modern equation rather than a function defined by *TMM*. As explained, this was felt to be justifiable because Flamsteed's method of deriving the Equation of Centre from elliptic orbits agreed within arcseconds of the modern formula (Ch. 6, III). Our computer model could be criticised for not properly modelling errors likely to arise at this step, from interpolating an Equation of Centre table, e.g. in Streete's *Astronomia Carolina*, as reprinted by Halley in 1710. It therefore tends to achieve undue accuracy as compared to historical worked examples, as will appear later—that is, the model works too well.

V. Amplitude of the Variation

The Variation was one of the three known lunar inequalities. The Moon moved more swiftly at syzygy than at quadrature, causing the Variation to reach its maximum "equation" in the octants. In Proposition 66 of Book One of *PNPM* Newton undertook to give a derivation of it, as resulting from the Sun's pull on the lunar orbit. Tycho Brahe had announced its discovery in 1598, giving it an amplitude of 40'.5, which was quite close to its true value of 39'.5. Horrocks in his first exposition in the 1630s had settled on a smaller value of 36' 45", which he later reduced to 36' 27" (Horrocks 1673, p. 483). Despite being a keen disciple of Tycho Brahe's colleague Kepler, Horrocks adopted this smaller value for his Variation term. In *TMM* Newton reduced it to a yet smaller value of 35' 32". His letter to Flamsteed of November 1st, 1694, discussed this matter.

This divergence baffled commentators, with the nineteenth-century astronomer Gaythorpe declaring that the British astronomers had been simply mistaken (Gaythorpe 1956, p. 40). Only recently was it discovered by Jorgensen (1974, p. 317) that the Horrocksian mechanism incorporated a sizeable fraction of the Variation, in fact "some 5'.25 of the variation" (*GHA*, p. 265), so that its correct amplitude in the Horrocksian theory is a mere

34′ 15″. The term used by Newton and Halley was thus more or less correct.

The *GHA* averred that Flamsteed took a value for the Variation of 36′ 45″ "obtained on the basis of observation" (*GHA*, p. 264). To find Flamsteed's value for the Variation, we inspect *DOS's* table for Variation, finding that it peaks at 38′, the same value as was later adopted by William Whiston in his 1707 opus (in Whiston 1715, p. 34).

TMM makes the term vary with the seasons:

> **The greatest Variation of the Moon (***viz***, that which happens when the Moon is in an Octant with the Sun) is, nearly, reciprocally as the cube of the Distance of the Sun from the Earth. Let that be taken as 37′.25″. when the Sun is *in Perigeo*, and 33′.40″. when he is *in Apogeo*: and let the Differences of this Variation in the Octants be made reciprocally as the Cube of the Distances of the Sun from the Earth; and so let a Table be made of the aforesaid Variation of the Moon in her *Octants* (or its Logarithms) to every *Tenth, Sixth,* or Fifth Degree of the mean Anomaly: And for the Variation out of the Octants, make, as Radius to the Sine of the double Distance of the Moon from the next Syzygy or Quadrature : : [sic] so let the aforefound Variation in the Octant be to the Variation congruous to any other Aspect; and this *added* to the Moon's place before found in the first and third Quadrant (accounting from the Sun) or *subducted* from it in the second and fourth, will give the Moon's Place equated a fifth time.**

TMM here specifies a sine $2(M - S)$ function, with two cycles per lunar month, zero at the four quarters, and an amplitude of $35\frac{1}{2}′$. This varies with the seasons, being maximal at perihelion (what *TMM* calls the Sun *in Perigeo*) and minimum at aphelion, which we model using a cosine function.

As once before, an inverse cube relation to the solar distance is affirmed, as can be viewed as implying an inverse-square gravity principle, discussed in the Second Edition of *PNPM*. This would imply a ten percent fluctuation in the course of the year. Thereby our fifth equation becomes:

$$M_5 = M_4 + \left[0.592 - 0.031 \cos (H - S_1)\right] \times \sin 2(M_4 - S_1)$$

The small value of Variation gives a clear means of checking whether astronomers were using a Horrocks-based system.

Next we come to the sixth equation, which marred *TMM* in its initial version, as it was given the wrong way round. In the next chapter we see how Newton's additional equations for *TMM*'s seven steps were all valid, except that this one operated in reverse, adding where it should be subtracted, as he later realised. Having a correct equation the wrong way round is much worse than having an irrelevant or mistaken equation: it continually creates an error of twice its amplitude. Here are the directions for the sixth and seventh:

> **Again, as Radius to the Sine of the Sum of the Distances of the Moon from the Sun, and of her Apogee from the Sun's Apogee (or the sine of the Excess of that sum above 360°.) : : so is 2′. 10″. to a sixth Equation of the Moon's Place, which must be *subtracted*, if the aforesaid Sum or Excess be less than a Semicircle, but *added*, if it be greater. Let it be made also, as Radius to the Sine of the Moon's Distance from the Sun : : so 2′. 20″ to a seventh Equation: which, when the Moon's light is encreasing, *add*, but when decreasing, *subtract*; and the Moon's place will be equated a seventh time, and this is her place *in her proper Orbit*."**

The expression "::" meant "in proportion to," used for comparing or equating (in the modern sense) ratios. Effectively, we are instructed to sum $(S - M)$ and $(H - A)$ in a sine function having an amplitude of 2′ 10″, which gives:

$$M_6 = M_5 + 0.0361 \sin (S - M + H - A)$$

William Whiston's comment upon the sixth, made in his Lucasian lectures to students of Cambridge University in 1703, is often quoted:

> How it should come to pass that this sixth Equation of the Moon should arise from Causes which are so unlike join'd together amongst themselves, as are the motion of the Moon from the Sun, and the Motion of the Apogee of the Moon from the Apogee of the Sun; I must acknowledge myself altogether ignorant; nor is there Opportunity for enquiring in these Matters merely Astronomical. In the mean while, I suspect that this Equation was rather deduced from Mr *Flamsteed's* observations, than from Sir *Isaac Newton's own Argumentation.* (Cohen 1975, p. 361)

We will shortly observe how the modern equations of lunar longitude, at this amplitude range of 2–3′, contain much stranger-looking combinations of terms than the above, as puzzled Mr Whiston.

Finding the Moon's place "in her proper Orbit" would seem to imply that all the above computations have not been in the plane of the ecliptic, but rather in an orbit tilted at five degrees thereto. There are certain difficulties in such a view, which, fortunately, need not concern us. *TMM* does not give instructions over the "reduction," for converting to ecliptic longitude, this being a standard procedure.

The seventh equation varies with lunar phase, additive in the waxing period and subtractive in the waning, i.e., as sin $(S - M)$. Its amplitude is modulated by a nine-year period:

> **Note here, the Equation thus produced by the mean Quantity 2′. 20″. is not always of the same Magnitude, but is encreased and diminished according to the Position of the Lunar Apogee. For if the Moon's Apogee be in conjunction with the sun's, the aforesaid Equation is about 54″. greater: but when the Apogees are in opposition, ˋtis about as much less; and it librates between its greatest Quantity 3′.14″. and its least 1′.26″. And this is when the Lunar Apogee is in Conjunction or Opposition with the Sun's: But in the Quadratures the aforesaid Equation is to be lessened about 50″. or one minute, when the Apogees of the Sun and Moon are in Conjunction; but if they are in Opposition, for want of a sufficient number of Observations, I cannot determine whether it is to be lessen'd or increas'd. And even as to the Argument or Decrement of the Equation 2′.20″. above mentioned, I dare determine nothing certain, for the same reason, *viz* the want of Observation accurately made.**

This is a cosine function because it reaches a maximum when solar and lunar apses align, i.e., perigee conjoins perihelion, so we represent it as cos $(H - A_2)$. The apse-position twice-equated is used:

$$M_7 = M_6 + 0.0389[1 + 0.3857 \cos (H - A_2)] \times \sin (S_1 - M_6).$$

At waxing Moon, lunar longitude must be larger than solar longitude, so the function sin $(S - M)$ must be negative, contrary to the above-quoted instructions. Professor Wilson, however, in *GHA* (p. 267), quoted the seventh equation in the above form. In correspondence, he pointed out that the Latin text of Gregory's opus, published in 1702, has the converse instruction, namely "Hanc aufer quando Lunae Lumen augetur, & (e contra) adde cum illud minuitur." (Cohen 1975, p. 127). Likewise, an English translation of Gregory's text, published later in 1715, also reproduced in the Cohen opus, gives that version, which we should presumably take as authentic. In addition, this is the correct sense in the modern equation. It remains hard to imagine Edmond Halley, if indeed he was the producer of the *TMM* version we have been using, and possibly its translator, introducing such an error. The matter remains conjectural.

There is the hint of a second modulation to the seventh, in the reference to Conjunction and Opposition, which we have ignored. A further modulation of both sixth and seventh equations follows, which can also be ignored as well below the limit of detectability, adjusting by a mere few percent the anomalistic cycle:

> **If the sixth and seventh Equations are augmented or diminished in a reciprocal *Ratio* of the Distance of the Moon from the Earth, *i.e.*, in a direct *Ratio* of the Moon's Horizontal Parallax; they will become more accurate: And this may readily be done, if Tables are first made to each Minute of the said Parallax, and to every sixth or fifth Degree of the Argument of the sixth Equation for the *Sixth*, as of the Distance of the Moon from the Sun, for the Seventh Equation.**

There are two lunar nodes, the North or "ascending" and South or "descending," and *TMM* uses the first of these to define the Sun-node angle. The angle between Sun and lunar ascending-node is found, where both of these terms have had an 'annual equation' applied to them. They are both "equated" once, in *TMM*'s language, so the angle is $(N_1 - S_1)$. This angle undergoes one revolution over the "eclipse year" of 347 days, the interval between conjunctions of Sun and node. *TMM* instructs us to find this angle, in order to find the second node angle.

> **From the Sun's Place take the mean Motion of the Moon's ascending Node, equated as above; the Remainder shall be the Annual Argument of the Node, whence its second Equation may be computed after the following manner in the preceding Figure.**

VI. A Second Horrox-Wheel

A second Horrox-wheel now appears, for an "Annual Argument" of the nodes. This varies with the $6\frac{1}{2}$ month period, and it is all rather similar. A second diagram appears, identical with that here reproduced in section IV.

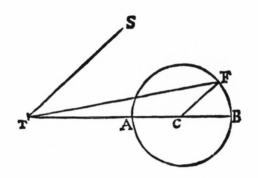

Let *T* as before represent the Earth, *TS* a right line conjoining the Earth and Sun: Let also the Line *TACB* be drawn to the Place of the Ascending Node of the moon, as above equated; and let *STA* be the Annual Argument of the Node. Take *TA* from a Scale, and let it be to *AB* : : as 56 to 3, or as 18 2/3 to 1, Then bisect *BA* in *C*, and on *C* as a Centre, with the distance *CA*, describe a Circle as *AFB*, and make the Angle *BCF* equal to double the Annual Argument of the Node before found: So shall the Angle *BTF* be the second Equation of the Ascending Node: which must be *added* when the Node is passing from a Quadrature to a Syzygy with the Sun, and *subducted* when the Node moves from a Syzygy towards a Quadrature. By which means the true Place of the node of the Lunar Orbit will be gained : whence from Tables made after the common way, the *Moon's latitude, and the reduction of her orbit to the Ecliptick,* may be computed, supposing the Inclination of the Moon's Orbit to the Ecliptic to be 4.59′.35″. when the Nodes are in quadrature with the Sun; and 5.17′.20″. when they are in Syzygys.

The ratio of *TA* to *AB* is $18\frac{2}{3}$ to 1, strangely echoing the period of the rotation of the nodes of 18.6 years. Our function *j* finds the angle *FTA*, the second nodal Equation, by dropping a perpendicular from *F* onto *TB* and taking the tangent of *FTC*.

The angle *STA* is $(N_1 - S_1)$, so the angle *FCB* being double its magnitude is $2(N_1 - S_1)$. The ratio of *TC* : *CB* is, by the ratio given above, 38.3:1. If the perpendicular is *FD*, then the tangent of *FTC* is $FD/(CD + TC)$, whence the function *j* is obtained. The maximum value of this angle is arcsin(1/38.3) or

1° 29′ 44″ (See fig. 7.4b). This second nodal equation appears *after* the seven steps of equation, being used solely for finding the reduction and celestial latitude.

Flamsteed's tables had a similar but larger node equation, of 1° 39′ 46″ at 45° anomaly; which equation came from Kepler (1627, p. 87) as Gaythorpe pointed out (1956, p. 142), and these node tables were used by Whiston (1715, p. 35). This Keplerian value is more accurate, i.e., nearer the modern value, than that of *TMM*. On the other hand, the amplitude of *TMM*'s newly-invented first nodal equation was, remarkably, within 2% of the modern value. Summarising, we may compare the maximal values of the two node equations as follows:

	Function	Modern value	TMM	Flamsteed/Kepler
First node equation:	$\sin (S - H)$	9′ 43″	9′ 30″	-
Second node equation:	$\sin 2 (S - N)$	1° 36′ 11″	1° 29′ 44″	1° 39′ 46″

(The two modern values were kindly obtained by Bernard Yallop, using Brower and Clemence's *Methods of Celestial Mechanics* (1961, p. 312).)

The amplitude of *TMM*'s second node equation is unspecified in the text, and had to be found by those constructing *TMM*-based tables. Persons composing such independently would be unlikely to arrive at the same magnitude to an arcsecond. This provides a "fingerprint" technique for showing any copying of data between table-makers.

The reduction as the final step for our *TMM* program was modelled on the reduction tables of *DOS*, but using the parameters given by *TMM*. The correction is zero at the two nodes, and also at their quadratures, i.e., it is a sin 2Θ function. The angle of the lunar orbit to the ecliptic varies, *TMM* tells us, as the Sun's angle to the nodal axis, being maximal at conjunction and minimal at quadrature. Consulting a standard table of reductions, for which we select Flamsteed's in *DOS* as was reprinted without alteration by LeMonnier in Paris in 1746, we observe that the reduction's maximal value varies between 6′33″ and 7′22″, depending on the orbit's angle of inclination. *TMM* instructs us to follow the tables prepared "after the common way." (p. 12, above). *DOS*'s maximal reduction varies by ±12%, while the angle of inclination of the orbit to the ecliptic is given by *TMM* as varying by merely ±6%, so the reduction is changing by twice as much as does the orbit angle each year. We require a modulating term with an amplitude of six percent, and the reduction term is therefore

$$M_{end} = M_7 + 0.116 \times \sin 2(N_2 - M_6)[1 + 0.059 \cos 2(N_2 - S_1)].$$

The cosine term gives maximal values for solar conjunctions with the nodal axis and smallest at their quadratures.

VII. Latitude

Celestial latitude varies as distance from the nodes, going through one cycle per draconic month, where longitude is measured from the North Node. Latitude is maximal at the quadrature position midway between the two nodes, and this maximum value is ± 5° 17′ 20″ when the nodes are conjunct with the Sun and ± 4° 59′ 35″ when in quadrature; thus there is a modulating function varying as 2Θ function, as for the reduction but with half the amplitude. The mean value is 5° 8′ 31″ or 5.142, a mere 16 arcseconds less than the modern value. Latitude becomes positive as the Moon passes the North node, when $(M - N)$ has a positive value. Putting the slower-moving position first, a sin $(N - M)$ function will require a minus sign in front. Thus, *TMM*'s instructions give us a celestial latitude formula of:

$$\text{Latitude} = -5.142 \times \sin (N_2 - M_6)[1 + 0.0288 \cos 2(N_2 - S_1)]°$$

Later on we ascertain how well this latitude function performs. Flamsteed averred that it was *TMM*'s weakest point in 1703: "The errors in latitude are frequently 2, 3, or 4 minutes, which is intolerable." (to Caswell, 23 March 1703, Baily 1837, p. 213).

TMM concludes with some remarks about parallax and refraction which, though relevant in practice, need not concern us. Overall, as far as monthly cycles are concerned, *TMM* appears as largely based on the tropical and anomalistic cycles, with the phase and nodal cycles only playing minor, ancillary roles.

VIII. An Early Draft of *TMM*?

A manuscript of Newton's published in the *Correspondence* (pp. 3–5, Volume IV) is entitled *A Theory of the Moon*. The commentary there stated that while there was "no clue" as to the date of its composition, it was probably written "some time prior to" *TMM*, adding that its text was published "almost verbatim" in *PNPM* of 1713.

Westfall referred to this manuscript in his view that:

A paper called "A Theory of the Moon" listed rules for computing seven corrections without discussing their theoretical foundation.... Several years later, Newton allowed David Gregory to take a copy of it and to publish it. (Westfall 1980, p. 547) adding that the version published in *Correspondence* was "probably from 1695." In Chapter Three, a composition date of *TMM* was suggested as 1700, i.e., shortly before Gregory saw it. Westfall's view, in contrast, is that Newton had virtually composed it some years earlier, and merely reproduced it in 1700. Of relevance here is an irate letter from Newton to Flamsteed of January 6th, 1699, when the latter had planned to mention, in a forthcoming opus of John Wallis, his contribution to Newton's endeavours over lunar theory. People were wondering what Flamsteed had been doing all these years as the "King's Observator," as he had published little, so he wanted to state his contribution towards the advance of theoretical astronomy, as having supplied the observations. After all, several years earlier he had heard stories, put about by Halley, that Newton had so far perfected the lunar theory that further observations by him were hardly necessary (Baily 1837, p. 162). Newton forbade this act, on the grounds that:

> with respect to the theory of the moon, I was concerned to be
> publicly brought upon the stage about what, perhaps, will
> never be fitted for the public, and thereby the world put into
> an expectation of what, perhaps, they are never likely to have.
> (6 Jan 1699, *Corr.* IV p. 296)

That must surely be read as an admission of failure, as a statement to his colleague that his endeavours had not been such as to warrant any proclamation to the learned world. Are we to believe that *TMM* had then been substantially composed, implying that the above-quoted words to Flamsteed were mere deception? Much depends here on whether *TMM* is viewed as having been a success, or a failure. The leading British theoretical astronomers, Halley, Whiston and Gregory, were as we have seen in no doubt upon this issue, once they saw it. That is why it has here been affirmed, that *TMM* was composed *after* the above-quoted remark and not before.

Westfall has claimed that *TMM* was composed in the 1690s, thereby by implication viewing it as a failure, since, as he rightly observed of this period, "Newton himself regarded the effort as a failure" (Westfall 1980, p. 547). Science historians, as was pointed out in Chapter One in discussing Bernard Cohen's view, have not generally viewed *TMM* as a working mechanism, that would define five positions in ecliptic longitude (Sun, Moon, lunar nodes and apse and aphelion) and one in latitude for any given time.

A letter from Newton to Halley of March 14, 1695 requested that the latter would deny prevalent reports that he was "about the longitude at sea." As this goal was the stated purpose of *TMM* when it was published, we must assume that no such composition had been formulated at the end of his main period of endeavour over lunar theory in the 1690s; to do so would imply a level of duplicity that we should not readily contemplate.

The view here taken, is that Newton did indeed regard his endeavours of the 1690s as a failure, but that he was then attempting to accomplish a derivation of the lunar inequalities from a gravity theory. Only after that had ended in failure, was *TMM* composed, effectively lacking reference to a gravity theory and merely improving Horrocks's kinematic model.

Against the Westfall thesis, let us note that the brief "earlier" manuscript has no seven rules or stages as does *TMM*, has no Horrocksian mechanism, is far from being a complete procedure for locating longitude positions, and is rather a fragmentary discussion of gravity theory as was attempted for the second edition of *PNPM*. I query the whole notion that it is an earlier draft. It refers merely to the first three equations of *TMM*, and none of the subsequent ones. Later on we address the manner in which gravity theory was related to the instructions of *TMM*, a matter of the utmost importance to subsequent astronomers, where a discussion of this manuscript's gravity arguments will be appropriate.

Comparing the two new annual equations introduced by *TMM*, viz. 20′ for the apse line and 9′ 30″ for the nodes, with those given in the unpublished "Theory," the latter are seen to be more exact. Its figures are 20′ 9″ for apse and 9′ 34″ for the nodes. An additional order of magnitude accuracy has appeared. There is no more distinctive difference between the first and second editions of the *Principia* than the increase in numerical accuracy. Often, the accuracy of the Second Edition went beyond what was warranted by the data, as if endeavouring to convey credibility by an increase in the number of decimal places (as Westfall described in his *Newton and the Fudge Factor* of 1973)(Westfall 1973). This strongly indicates, I suggest, the arrow of time, demonstrating that the undated manuscript was composed long after *TMM*, and not earlier as Westfall has assumed.

The manuscript has an interesting remark about the "annual equation":

> Moreover in deducing celestial motions from the laws of gravity we also discovered that the annual equation of the Moon's mean motion which Kepler and Horrocks coupled with the equation of time, but Flamsteed published separately, arises from the varying expansion of the Moon's orbit by the force

of the sun, in accordance with Corollary 6 to Proposition 66 in
Book 1. (*Corr.* IV, p. 4)

Thus, this equation derives from Flamsteed's discovery that Earth's rotation rate was constant, which he established from daily transits of Sirius. This was the major theoretical issue on which Flamsteed disagreed with Kepler, who had accounted for the annual equation by supposing that Earth's rotation rate varied over the course of the year. Subsequent astronomers credited Flamsteed with having discovered the Equation of Time, linking mean and apparent solar time, as the seasonal variation in day-length. In the *Principia*'s Second Edition, this passage appears with the reference to Flamsteed deleted, for reasons into which we need not enter. The maximal value of this equation here given, 11′ 55″, is in excess by about 43″.

The last paragraph of this manuscript (*Corr.* IV, p. 5) speculates about a nine-year cycle varying with the apse rotation (which would yield a $\sin(A - H)$ function). *TMM* has no period of this length—though Newton speculated upon one for the fourth equation as we saw, while the apse line crossed the solstice positions at zero Cancer-Capricorn. It lacks any terms longer than a Horroxian year of 411 days, a surprising feature from a modern viewpoint. Newton only grappled with the subject for just over half a year, from September 1694 until June 1695, and his most reliable positional data came from after 1690, when Flamsteed's mural arc was working. In contrast with this emphasis upon short-term cycles, Halley was convinced that the 18-year Saros cycle was of vital relevance. After their discussions on this topic, Newton may have considered incorporating a longer cycle into his theory:

> I have learned furthermore from the same theory of gravity that the Sun acts upon the Moon more strongly in the individual years when the Moon's apogee and the Sun's perigee are in conjunction than when they are in opposition. From this there arise two periodic equations, one for the Moon's mean motion, the other for the motion of her apogee. These equations are nil when the Moon's apogee is either in conjunction with the Sun's perigee, or in opposition to it, and in other positions of the apogee they have a given proportion to each other. The sum of these equations, when they are at their maximum, is about 19 or 20 minutes. (*Corr.* IV, p. 5)

This is a further basis for believing that it was composed years after *TMM*, perhaps a decade or so later.

While having to disagree with both Westfall and the *Correspondence* commentator, our conclusion happily accords with Whiteside's view: he characterised this published manuscript as an "initial version" of the opening paragraphs of the revised scholium of Proposition 35 of Book Three of the 1713 Second Edition of *PNPM* (Whiteside 1976, p. 327, Note 46).

Chapter 10

Testing the Sevenfold Chain

Having formulated a model in accord with Newton's instructions, we now test its validity. There are three ways we will do this, the first being a comparison with historic computations by astronomers who adopted the *TMM* procedures. Such a comparison may help to establish confidence in the validity of the TMM-PC model; and further, to ascertain the extent to which historical authors used *TMM*, a question not yet well resolved by historians.

The other two approaches to be developed in this chapter test the individual steps of *TMM*. This can be done theoretically, by comparing modern equations of lunar longitude with those of *TMM*. Thereby we evaluate statements made by Baily, Whiteside and Wilson on the subject. Complementing this is a practical approach, whereby any step of *TMM* can be tested by altering TMM-PC in some respect, and noting whether, on average, it thereby becomes more or less accurate.

The latter method should be able to give a definite answer as to the validity of any component of *TMM*, as may not be readily discerned from theoretical considerations. After all, none of the modern equations have their amplitudes modulated, in the way of *TMM*, by long-period functions. If, for example, we should be curious as to how much of an improvement was accomplished by Halley's modification of the Horrocksian model for eccentricity, as compared with Flamsteed's model, then such a testing on TMM-PC should resolve the matter.

I. Five Historic Case-Studies

Astronomy textbooks of the period always gave examples of longitude computation. TMM-PC can model these worked examples, provided their method was Newtonian. We endeavour to avoid worked examples involving eclipses, since exact longitude would then be known, as solar longitude could be determined with great accuracy. These would provide a tempting opportunity for the astronomer to claim a greater accuracy. By the time that the worked examples appear in the 1730s and 1740s a systematic error had accrued of nearly two arcminutes in the mean lunar motion.

If, as with Halley and Leadbetter, the method involved logarithms, the steps are less easy to follow. The present approach overcomes this difficulty, by viewing merely the beginning and end of the operational sequence. For the selection of our case-studies, we are guided by the most recent claim as regards who adopted *TMM*'s procedures, made by Professor Wilson in *GHA*:

> Newton's rules for calculating the place of the Moon were incorporated into the tables of Charles Leadbeater's *Uranoscopia* (1735); in the tables that Flamsteed constructed about 1702 and which, having been given by Halley to P. C. Lemonnier, were published by the latter in his *Institutions astronomiques* of 1746; and in Halley's *Tabulae astronomicae* (1749). (*GHA*, p. 267)

William Whewell gave, in the nineteenth century, a more extensive list of such persons, which a later chapter will consider. Our historical comparison will use the works above-cited by Wilson, plus the Dunthorne example treated earlier.

Table 10.1 gives the "mean moon" position for the local mean time in the left hand column, as the starting-point of the historic computation, and below it is the answer given in the textbooks, as final ecliptic longitude. To the right of these historic computations are those of TMM-PC for these same times, showing the equivalent values it generates for that moment in time. Subtraction gives the difference between the two methods, in the "$a-b$" column.

Below each of these is a modern (ILE) estimate of the true longitude for that instant, in italics, plus the errors of the "Newtonian" method by comparison with that value, both of the historic textbooks and of our TMM-PC program, cited in arcminutes. All longitudes have been converted to degrees, thus what was then given as 6s 27° 59' 18" becomes 207.988. In the first example, Leadbetter can be seen to agree with the TMM-PC mean value within ten arcseconds, an acceptable error for him to make in reading his tables.

We have seen how Dunthorne is the case-study which exactly mirrors TMM-PC, and his answer differs from TMM-PC by a mere 0.001° = 4". The others have generally made slight adjustments, chiefly to the sixth and seventh equations, either omitting them or reversing the sign of the former. The others, chiefly Leadbetter, Halley and LeMonnier, reversed the sign of the 6th equation as in the 1713 version. As it has up to two and a half minutes amplitude, this is of considerable significance.

Table 10.1: *TMM* **Computations, Historic vs. Computer, of Lunar Longitude**

		Historic degrees a	*TMM-PC* degrees b	*Difference* arcseconds a − b	
Leadbetter (1735)					
May 7th 1731, 10hrs	mean:	207.988	207.991	−10″	
	answer:	202.337	202.337	+0″	
actual position:			202.407		
errors (answer−true posn):		−4′.2	−4′.2		
Leadbetter (1735)					
Sept 16 1734 noon	mean:	183.112	183.111	+3″	
	answer:	188.426	188.424	+6″	
	actual:		188.459		
	errors:	−2′.0	−2′.1		
Dunthorne (1739)					
Jan 2nd 1737, 3h 40m	mean:	80.119	80.119	0″	
	answer:	74.137	74.136	+ 4″	− 6th
	actual:		74.251		
	errors:	−6′.8	−6′.9		
LeMonnier (1746)					
Aug 4 1739, 5h 55m 25s	mean:	134.869	134.862	+25″	
	answer:	132.642	132.639	+9″	
	actual:		132.622		
	errors:	+1′.2	+1′.0		
Halley (1749)					
Dec 5th 1725, 9h 8m 5s	mean:	51.428	51.431	− 9″	
	answer:	45.709	45.701	+25″	
	actual:		45.713		
	errors:	−0′.2	−0′.7		

The agreement in the right-hand column is generally within arcseconds, thereby establishing the *GHA's* view that the above persons were using *TMM*, albeit modified somewhat in the last two equations. It also supports Baily's rather cautious view that:

> It was not until the year 1735, when Leadbetter published his *Uranoscopia*, that we find a more perfect adoption of Gregory's *Newtonian rules* reduced to a tabular form" (Baily 1837, p. 702).

The modern longitude program, whose values have been shown in italics, runs on Julian time, and so is suitable for all five dates except that of LeMonnier in Paris. For LeMonnier's date and time, the three-stage conversion (Ch. 5, IV) involved subtracting eleven days, adding twelve hours to the given time, then subtracting nine minutes and twenty seconds to convert from Paris local time to G.M.T. This gave July 24th 1739 O.S., at 17 hours 46 minutes 5 seconds G.M.T., which was inserted into the program (The time-values fed into the *TMM* program for days after noon, December 31st 1680, GMT, Old Style were, respectively: 20456.153 (Dunthorne), 18389.417 (Leadbetter 1), 19617.000 (Leadbetter 2), 21389.240 (LeMonnier) and 16410.381 (Halley)).

We saw in chapter Five that Le Monnier's mean moon was more accurate than *TMM's* over this period. Using Le Monnier's tables, for the epoch date 1720, differences were compared from true values as in Chapter Five, showing that the mean position was then out by a mere 7 arcseconds. The same was done for the *TMM* program, whence we find that Lemonnier's mean moon was displaced 1' 12" or $0°.02$ from that of *TMM*: that amount was added to the first step of *TMM's* procedure solely for the LeMonnier example. Summarising, Le Monnier's operation sequences were modelled using TMM-PC with its sixth equation reversed, i.e., of correct sign, and with just over one arcminute added to its mean moon.

After that adjustment to LeMonnier's mean moon, his method still diverged from TMM-PC by forty arcseconds, which divergence arose in his fourth step, the Equation of Centre. The two worked examples by Charles Leadbetter had their first three equations echo *TMM*, but then, following Halley, *TMM's* sixth equation came next as the fourth, followed by the Equation of Centre. He also used Halley's mean motions (Appendix III) but differed in keeping the seventh equation. These two worked examples of Leadbetter's *Uranoscopia* of 1735 had values for the "Sun's true place" agreeing with *TMM* within one or two arcseconds.

As was earlier explained, the error in mean positions was almost two arcminutes for this period, and the error values can be seen to cluster around this value.

II. The Erronous Sixth

In the year 1713, Flamsteed wrote to a friend:

> I told you that the heavens rejected that equation of Sir
> I. Newton, which Gregory and Whiston called his sixth. I had
> then compared but 72 of my observations with the tables: now,
> I have examined above 100 more. I find them all firm in the
> same, and in the seventh too. And whereas Sir I. Newton has
> in his new book (pages 424 and 425) thrown off his sixth, and
> introduced one of near the same bigness but always of a con-
> trary denomination, and a bigger in the room of the seventh, if
> I reject them both, the numbers will agree something better
> with the heavens than if I retain them: so that I have deter-
> mined to lay these *crotchets* of Sir I. Newton's wholly aside.
> (Baily 1837, p. 698; cf. *PNPM* pp. 662–3)

This view of Flamsteed's appeared in response to the new edition of *PNPM* (Baily 1837, p. 698). Earlier he had commented in general terms about discarding some of *TMM*'s ancillary equations, but this was his first definite statement upon the matter. He has plainly noticed the reversal of sign for the sixth and enlargement of amplitude of both sixth and seventh, but was still not impressed. His unfortunate conclusion was, that both the sixth and seventh equations were best omitted, and that "the heavens rejected" the sixth even with its sign reversed.

Flamsteed was probably the first to discern the erroneous sign for the sixth equation, but otherwise his view is mistaken. For, as we shall now show, all four of the Newtonian ancillary equations turn up in the modern formulae. It is ironic that the person who supplied the data from which the theory was wrought, should end up sceptical about what had been attained. Before making such a comparison, our versions of the formulae are compared with those of others.

III. Newton's New Equations

Four new equations appeared as the second, third, sixth and seventh stages of *TMM*. The present work is the fourth to propose an algebraic format for them. Versions given previously by Baily (1837), Whiteside (1975) and Wilson (1989) are here compared with ours, omitting the amplitude-modulating terms. Symbols used in the present text are employed for the comparisons. To avoid ambiguity over signs, we quote *GHA* that "$M - S$ is the angular distance of the Moon from the Sun" (p. 267): envisaging the luminaries as revolving anticlockwise around Earth, angles measured anticlockwise are positive.

The second Equation:

Baily	$3'\,45'' \sin 2(A - S)$	given as $\sin 2[(M - S) - (M - A)]$
Whiteside	$3'\,45'' \sin 2(S - A)$	
Wilson (GHA)	$3'\,45'' \sin 2(A - S)$	given as $-\sin 2(S - A)$
TMM-PC	$3\,'45'' \sin 2(A - S)$	

Whiteside's term is reversed in sign as compared with the other three.

Third equation:

Baily	$47'' \sin 2(S - N)$	
Whiteside	$47'' \sin 2(S - N)$	
Wilson	$47'' \sin 2(N - S)$	given as $-\sin 2(S - N)$
TMM-PC	$47'' \sin 2(N - S)$	

Both Baily and Whiteside have the functions in reverse mode, i.e., 180° out of phase as compared with Wilson. I ascertained the plus and minus signs largely empirically, by writing the equation into the computer then observing whether the plus/minus values changed in accord with *TMM*'s instructions for varying time-values.

Sixth equation:

Baily	$2'\,10'' \sin (M - S + A - H)$
	given as $\sin[2(M - S) + (S - H) - (M - A)]$
Whiteside	$2'\,10'' \sin (M - S + A - H)$
Wilson	$2'\,10'' \sin (S - M + H - A)$
	given as $-\sin (M - S + A - H)$
TMM-PC	$2'\,10'' \sin (S - M + H - A)$

The first two have the signs reversed as compared with the others.

Seventh Equation:

Baily	$2'\,20'' \sin (M - S)$	
Whiteside	$2'\,20'' \sin (M - S)$	
Wilson	$2'\,20'' \sin (S - M)$	given as $-\sin (M - S)$
TMM-PC	$2'\,20'' \sin (S - M)$	

In a sense, Baily and Whiteside were unconcerned with the signs of these terms, which only have meaning within an operating system.

IV. Comparison with Modern Terms

The new equations of *TMM* appeared, as Baily complained, without justification:

> Newton has not explained, in the document under review, how he deduced these new equations, nor whether any of them are derived from his own theory of gravitation, or from Horrox' theory of the libratory motion of the lunar apogee. (Baily 1837, p. 694)

While this is true, it will here be argued, in contrast with the views of others on this matter, that the new equations showed the profound intuition of their author. Not only did Newton originate the concept of ancillary equations in this context, an unheard-of thing prior to about 1695–6, but all four of them were substantially valid, and even had near to their optimal amplitudes. *TMM* was marred by having its sixth equation the wrong way round, however this was corrected in 1713, well prior to the period when astronomers commenced using it. We have already seen how several astronomers who took up the Newtonian theory accomplished this vital reversal of sign in the sixth equation.

The modern equations for lunar longitude are normally cast in terms of just four terms: solar anomaly (M), lunar anomaly (M'), lunar elongation (D) (angular distance from the Sun) and mean lunar distance from ascending node (F). These are used because they turn up most often in the hundreds of terms comprising the theory. We may add an asterisk to the modern solar anomaly term, as M^*, to avoid confusion with the *TMM* symbol. Before comparing these with the *TMM* program, we should recall that the modern definitions of anomaly, with respect to perigee and perihelion, are 180° out of phase with the old. We then transform them using the symbols M (Moon), S (Sun), N (node), A (Apogee), and H (Aphelion); thus,

$$D = M - S,$$

$$F = M - N,$$

$$M^* = S - H + 180°,$$

and

$$M' = M - A + 180°.$$

For example, the eleventh of the modern terms in longitude

$$+0.041 \sin (M' - M^*)$$

Modern	Equivalent	Neutonian	
1) +6.2888 sin M'	6° 17' 24" sin (M − A)	ellipse function	
2) +1.274 sin (2D − M')	1° 16' 26" sin (M + A − 2S − 180)	evection	
3) +0.658 sin 2D	39' 29" sin 2(M − S)	35' 32" sin 2(M − S)	5th equation
4) +0.213 sin 2M'	12' 49" sin 2(M − A)	Horrocks theory	
5) −0.185 sin M*	−11' 8" sin (S − H)	11' 49" sin (H − S)	1st equation
6) −0.114 sin 2F	−6' 51" sin 2(M − N)	6' 57" sin 2(N − M)	reduction
7) +0.058 sin (2D − 2M')	3' 32" sin 2(A − S)	3' 45" sin 2(A − S)	2nd equation
8) +0.057 sin (2D − M* − M')	3' 26" sin (M − 3S + A + H)	-	
9) +0.053 sin (2D + M')	3' 12" sin (3M − A − 2S)	-	
10) +0.046 sin (2D − M*)	−2' 44" sin (2M − 3S + H)	-	
11) +0.041 sin (M' − M*)	2' 28" sin (M − S + H − A)	−2' 25" sin (S − M + A − H)	6th, 1713
12) −0.034 sin D	−2' 5" sin (M − S)	2' 20" sin (S − M)	7th equation
13) −0.030 sin (M* + M')	−1' 49" sin (M − A + S − H)		
14) +0.015 sin (2D − 2F)	55" sin 2(N − S)	47" sin 2(N − S)	3rd equation

Table 10.2: The first fourteen modern terms for lunar longitude are given on the left, then restated in the adjacent column using TMM-PC symbols. The TMM equations are given in the third column, where the first equation refers to the annual equation and the fifth to the Variation.

becomes

$$2'\,28'' \sin \left[(M - A) - (S - H)\right]$$

or

$$2'\,28'' \sin (M - S + H - A),$$

which we can recognise as the sixth equation. The first fourteen modern equations in order of diminishing amplitude are as in Table 10.2, opposite.

Only to a limited extent can we compare these sets of terms, as their mode of use differs: the modern procedure uses the *same* mean values at every stage, whereas the Newtonian procedure changes these at each step by "equating" them. The first four terms indicate however the fundamentally different basis of modern lunar theory from that of Newton. With the annual equation, the fifth in the modern sequence, the first of *TMM*'s steps appears. It is evident that three or four of the modern terms do not correspond with anything in the old theory. If omitted, their amplitudes are such that they would generate errors well beyond the 6 arcminutes or so maximum of *TMM*.

Baily acknowledged the validity of three of the four new equations, though appearing very doubtful about their amplitude. Of the second, he wrote rather sceptically that: "This equation, depending on twice the annual argument, or $2(D - A)$ according to Delambre's system of notation, does not amount to so much as 1' in the tables of Mayer, Burgh or Burckhardt." (Baily 1837, p. 737). That is curious, since in the modern scheme it amounts to $3\frac{1}{2}$ arcminutes, i.e., it should have been larger.

On the small third equation, Baily commented that it was "somewhat greater in the tables of Mayer, Burgh, and Burckhardt." (Baily 1837, p. 737). One hopes it was not much greater, since in the modern scheme its amplitude is a mere 55". Of the sixth, Baily observed that its coefficient and sign as given in the 1713 *Principia* were adopted by Halley. (Baily 1837, p. 740).

Whiteside in his 1976 essay did not commit himself to affirming that any of the new, Newtonian equations were valid, but merely concluded: "Pity those—notably Halley—who in the early decades of the eighteenth century tried to found solidly accurate tables of the moon's motion upon such a flimsy, rickety basis." (Whiteside 1986, p. 324). That was far from being Halley's view. We are not able to support the *GHA* view that, of the four new equations, only the second "has the correct form and, very nearly, the correct coefficient" (*GHA* p. 267).

V. "Apogee in ye Summer Signs"

In the winter of 1694/5 the lunar and solar apses drew into alignment as
the Sun crossed over them both at midwinter. In November of that year
Newton sent an urgent request for lunar data when "apogee is in ye
summer signs" (*Corr.* IV p. 47). Newton sought in vain for any $(A - H)$
perturbation term linked to this nine-year cycle, absent from his *TMM*. On
July 27th 1695 he wrote to Flamsteed,

> I had rather you would send me those [observations] from
> Aug. 24th, 1685, to July 5th, 1686, when the aphelium was in
> the same position as in the year 1677. (Baily 1837, p. 159.)

At the Full Moon of December 21 1694, mean apogee was at 5° Cancer (a
"summer sign"), merely 5° away from the syzygy axis. It would appear
from the above equations that there are no simple terms involving the
$(A - H)$ function that could have been discovered.

Newton urged Flamsteed to take lunar observations because of the great
importance of "apogee in ye summer signs," during "ye sun's opposition in
midwinter." (Letter of Nov. 17, 1694, in Baily 1837, p. 140). From the *TMM*
program we discern that, when the Sun reached zero Capricorn at the midwinter solstice, the mean apogee stood within a degree or so of zero Cancer,
so they were in close opposition. Newton's next letter reiterated the urgency
of the matter:

> For the position of the apogee in the Sun's opposition in mid-
> winter is a case of great moment and will not return for many
> years. The observation in the full and both the quadratures are
> of greatest moment. (Letter of December 4th, 1694, Baily 1837,
> p. 143).

In earlier days, Flamsteed had been concerned to locate apogee by use of the
micrometer screw gauge on his telescope eyepiece to measure lunar diameter, a method he did much to pioneer, so he would have appreciated this
event. But, as to what equation Newton was then searching for, we remain
in the dark. Evidently, by the time *TMM* was composed, he had reached no
conclusion as to the relevance of the nine-year apse cycle, except for a
minor role in modulating the sixth equation. (For how this new technology
was developed in the North of England, chiefly by Yorkshireman William
Gascoigne, and how Flamsteed came to hear about it, see Chapman 1982,
p. 21, and Chapman 1990 pp. 35–7).

Would it have been possible for Newton to discern any further lunar
equations? One answer is, that no further equations were discernible at that

time, as they were too complex; that he found all there was to find, then left Cambridge for London. His sustained work on the lunar theory occupied half a year, from September '94 to June '95. In September of that year, Halley discovered the cyclic return of the comet that bears his name, which may have tended to move Newton's attention away from the subject. In the autumn of 1695 *TMM*'s author accepted a job as Master of the Mint: on November 26, 1695, Wallis wrote to Halley, "We are told here [Oxford] that he [Newton] is made Master of the Mint, which if so, I do congratulate him." (More 1934, p. 435).

Terms such as "sin $(3M - A - S)$" are not intelligible as the seventeenth and first half of the eighteenth-century understood the notion: they merely come out of the mathematics after complex differential equations have been applied, and accordingly belong to an entirely different epoch.

VI. Testing the New Equations

To investigate *TMM*'s four new equations, a sampling period of 160 days was chosen. This was selected as avoiding multiples and fractions of *TMM*'s main cycles, namely 365, 205, 29.5 and 27.5 days. Forty such positions following the epoch date of 1680 were generated on TMM-PC. Equivalent longitudes were generated on a modern program and imported into the spreadsheet where the two values were subtracted. Thereby the mean and standard deviations of a column of their differences were obtained. The means of these samples ought to be close to the error value of mean lunar motion used by *TMM*, or not significantly different from it, if our sample is indeed random with respect to the rhythms of the mechanism we are investigating.

The second, third and seventh equations were omitted, one at a time. This was done using *TMM*-2, i.e., with Newton's 1713 value of the sixth equation. As this version is more accurate, it will plainly be more sensitive to other factors. The third equation is of very small amplitude, so as can be seen its sign had to be reversed for appreciable effect. Lastly, the Flamsteed-Horrocks method of varying the eccentricity was used in place of the Halley-Newton method. The results, citing the percentage increases of standard error, are presented in Table 10.3, below.

The results shown as **b–e** in the Table confirm what was inferred from theoretical considerations above, that all four of the "new" Newtonian equations contribute to predictive accuracy. Both their removal and their sign reversal impaired *TMM*'s function—contrary to Flamsteed's opinion.

Table 10.3: Accuracies of *TMM*-2 Modifications, showing percentage error increases in right-hand column compared with that of *TMM*-2. E.g., on removal of the second equation in (b), the S.D. of errors appeared as ±3.08 arcminutes, this being 64% more than 1.88, its optimal value.

a TMM-PC-2:	−0.38 ± 1.88′	
b no second:	−0.43 ± 3.08′	64% more
c no third:	−0.40 ± 2.03′	8% more
d reversal of third:	−0.42 ± 2.30′	22% more
e no seventh:	−0.40 ± 2.75′	46% more
f TMM-PC (with wrong 6th):	−0.50 ± 3.77′	200% more
g *DOS* eccentricity:	−0.53 ± 4.34′	230% more.

Line **f** estimates the error in the original version of *TMM*, a result comparable to that found earlier (Ch. 8, IV).

Flamsteed's Horrocksian method for finding eccentricity left it reduced by a factor cos δ as Gaythorpe pointed out (Ch. 7, IV), where δ is the second equation of apogee. To estimate his method's accuracy, the *f* function in TMM-PC was replaced by a simpler, Horrocksian movement (Ch. 7, II). The rev. William Whewell took a derogatory view of Halley's adjustment in this regard, as being a mere "slight alteration" (Ch. 7, IV), and similarly Whiteside expressed the view that "Newton himself (not that it matters a great deal from a computational viewpoint) was in fact to adopt Halley's variant on this…" (Whiteside 1976, p. 327 note 35). In contrast, our quantitative approach has revealed that Halley's contribution was the greatest single improvement whereby the Newtonian approach differed from the Horrocksian. Two-thirds of the error was removed by that one adjustment (line **g** in Table 10.3), thereby confirming Newton's own view that it was a "very good" improvement (*Corr.* IV p. 34).

Next, an attempt was made to optimise the program, by giving the four new equations the amplitudes of their corresponding above-quoted modern functions. Those of the second and third were reduced slightly, while for the sixth and seventh they were increased. Curiously, this increased the standard deviation by one percent. Next, the lunar eccentricity value was decreased, from the 0.055050 value of *TMM*, to 0.05490 as the modern value for this constant (equivalent to using the modern equation's amplitude, as in the above-cited Table, of 6° 17′ 24″, in place of *TMM*'s 6° 18′ 3″). Again, a slight increase in standard deviation for the forty data-points was observed. I could not find any case where adjustment of the *TMM* parameters improved accuracy.

A statistical "confidence limit" is normally taken as double the standard deviation, being the range containing 95% of the data. On this basis, *TMM*-2 had a confidence limit of 3.6 arcminutes. A different view has been adopted by Sir Alan Cook on this matter, namely that a single standard deviation value is equivalent to what the historical characters gave as an error-limit. He has repeatedly expressed the view that Halley was correct in affirming that *TMM* (or Halley's version thereof) gave an accuracy within the much-desired two minutes of arc (Cook 1996, p. 351; 1997, p. 186; 1998, pp. 372, 375). The view here adopted is that it is inappropriate to take one standard deviation of the error generated by a theory as equivalent to a historically-expressed confidence limit: a mere sixty percent of values will fall within that range. Further, it may not well express Halley's view to state that he claimed a two-arcminute accuracy, such a view being more associated with David Gregory.

VII. Comparison with DOS

Reconstructing the *DOS* procedure will help us to appreciate both the relation between Flamsteed and Newton, and what was meant by "Horrocksian." Such a model ought to generate the same errors as Newton and Gregory were shown three centuries ago on their visit to Greenwich.

The idea of tackling lunar theory seems to have come to Newton in September of 1694 when with David Gregory he paid a visit to Flamsteed at Greenwich, and was shown a table of lunar latitudes and longitudes as observed, together with their discrepancies from what Flamsteed calculated ought to be their positions (both Gregory and Flamsteed have left notes of this event). The challenge was for Newton to construct something better than the *DOS* method.

Flamsteed's *De Sphaera* gave the furthest development of the Horrocksian method of finding lunar longitude. It is the proper point of comparison for assessing *TMM*, being its immediate ancestor. William Whiston, in his astronomy lectures to the Cambridge University mathematics students in the year 1703, advised them:

> Take, therefore, Mr Horrox's Lunar Hypothesis, as cultivated
> and explained by Mr Flamsteed." (Whiston 1716, p. 104)

The method had three stages: the annual equation, Equation of Centre, and Variation. They are similarly described in the procedure given by Flamsteed in Horrox's *Opera Omnia* of 1673 (Horrox 1673, p. 494), except for minor alterations in constants and mean motions.

We shall call the program simulating the *DOS* procedure DOS-PC. Flamsteed's *De Sphaera* dealt with many other issues, but "DOS-PC" designates solely its method of finding lunar longitude. The *TMM* program was deconstructed to reach this more primitive procedure, as follows: remove all of equations 2, 3, 6 and 7, and the annual equations from node and apse; diminish solar eccentricity to the Horroxian value; remove the modulating factor from equation 5 (the Variation) and increase its amplitude to 38 arcminutes; insert *DOS* epoch values in place of *TMM*'s (Ch. 5, II; also Appendix III); the daily mean lunar motion was also adjusted, the difference between *DOS* and *TMM* being one arcsecond a year; the *DOS* parameters for lunar eccentricity were used, measuring the latter by the function:

$$E = 0.05524 + 0.01162 \cos 2\Phi$$

in place of the Newtonian

$$E = 0.05505\sqrt{1.0454 + 0.4262 \cos 2\Phi} \qquad \text{(Ch. 7, II);}$$

add in a proportionality factor to make the equation of apogee, as produced by the Horrox-wheel, slightly smaller, of maximal amplitude $11° 47' 22''$ in place of *TMM*'s $12° 15'$ (from Figure 7.4a, *DOS*'s equation of apogee δ is given by $\tan \delta = FD/DT = 0.011286 \sin 2\Phi/E$, where E is the varying eccentricity. Figure 7.4b depicts the maximal value of δ which *DOS* gave as 11.789, where $CF/CT = \sin \delta = 0.2043$; taking the mean eccentricity line TC as equal to 0.055237, the radius FC is 0.011286); and simplify the second equation of the node, so that it becomes a sine function of $2(S_1 - N)$, without any Horrox-wheel mechanism as *TMM* gave it.

Flamsteed's *DOS* outlined the method in ten steps, of which the first comprised his "Equation of Days" of which he was regarded as the pioneer (see, e.g., Charles Leadbetter's comment in his *Uranoscopia* of 1735 (Leadbetter 1735, p. 2); whereby the uniform rotation of the Earth in sidereal space became the basis for the definition of time, mean time as opposed to clock time, later standardised as Greenwich Mean Time.

His steps two and three obtained the mean motions from tables. Step four subtracted out the annual equation for the two luminaries. Step five used what was called the "Annual Argument" and which we have called the Horrox-angle, to "equate" the apogee and eccentricity. The sixth used the mean anomaly (i.e., $A_1 - M_1$) to give the true equation of orbit; which we may represent by

$$M_2 = M_1 + h\{A_1 - M_1\},$$

where h is the equation of centre function. Then the seventh stage adds the Variation. As our TMM-PC used the *DOS* reduction procedure, no adjustments are here required. The node had to be once equated before it could be used for finding the reduction and latitude, using tables based on the $(S_1 - N)$ angle. The overall latitude angle, i.e., the tilt of the Moon's orbit, likewise varied with that angle. The node's maximal equation, i.e., the maximal value given in the *DOS* table, was 1° 40′. Thereby a program was

Flamsteed's Lunar Theory

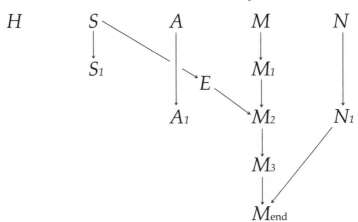

constructed, here shown in a flow diagram of its steps of equation, and then described by a sequence of trigonometric functions:

Mean Motions $M = 181.7328 + 13.1763946t$ $S = 290.580 + 0.9856469t$
$A = 244.1975 + 0.1114083t$ $H = 96.861 + 0.0000479t$
$N = 174.2430 - 0.0529550t,$

where t is the time in Julian days from noon of December 31st, 1680.

Annual Equation $M_1 = M - 0.197 \sin (H - S)$
$S_1 = S + 1.9368 \sin (H - S) - 0.0202 \sin 2(H - S)$

Equation of Centre $E = 0.055237 + 0.01162 \cos 2(A - S_1)$ (eccentricity)

$$A_1 = A - \arctan \frac{\sin 2(A - S_1)}{4.8943 + \cos 2(A - S_1)}$$

$$M_2 = M_1 + h\{E,$$

$(A_1 - M_1)\}$

where h is the equation of centre function (Ch. 6, II and Ch. 8, I).

Variation $M_3 = M_2 + 0.633 \sin 2(M_2 - S_1)$

Reduction $N_1 = N - 1.663 \sin 2(N - S_1)$

$$M_{\text{end}} = M_3 + 0.116 \sin 2(N_1 - M_3)[1 + 0.059 \cos 2(N_1 - S_1)]$$

There are, regrettably, only two worked examples whereby one can check this program. The worked example given in *DOS* has been discussed in Chapter Eight. In 1694 Flamsteed sent a table of computed positions to Newton, as will be analysed in the next chapter (Westfall 1980, p. 541; cf. p. 161 below). We might reasonably assume that he was using his earlier procedure, but his letter to Newton did not comment on this. William Whiston in 1703 used the same example as given in *DOS*; another was given by Cressner, discussed below. That is all, and it is not very much.

The *DOS* example had as we saw (Ch. 8, III, Table 8.3) an error of eleven arcminutes. London is five arcminutes due West of Greenwich, for which one-third of a minute was subtracted from the local mean time given to obtain GMT. The Table below compares the magnitudes of the three *DOS* equations: the annual equation, the Equation of Centre and the Variation, for the moment in the example of 6.35 p.m. London mean time, on 22 December 1680.

For $t = -8.72596$,

	DOS-PC	DOS	
Eccentricity:	57681	57678	parts per million
Apse equation:	10° 50′ 32″	10° 50′ 32″	
1) Annual eq.:	−1′ 03″	1′ 03″	
2) Eq. of centre:	−54′ 27″	−54′ 22″	
3) Variation:	−36′ 15″	−36′ 16″	
Reduction:	−3′ 59″	−4′ 00″	
Long. in ecliptic:	65° 9′ 51″	65° 9′ 52″	

Five arcseconds is the largest discrepancy in these steps of equation. This establishes that our program is correct. The accuracy of DOS-PC was then tested using the method described earlier, using forty positions generated at 160-day intervals. This gave a mean error of:

$$-2.4 \pm 6.5 \text{ arcminutes},$$

or a confidence limit of thirteen arcminutes. This well accords with Newton's remark made in a letter to Flamsteed of July 20, 1695, that "The Horrocksian theory...never errs above 10 or 12 minutes," (to Flamsteed, Baily 1837, p. 158); although, on January 15th, 1681, DOS-PC erred by 15 arcminutes. It also echoes the conclusion that Flamsteed expressed in the *Philosophical Transactions* of 1683, that "even the best" lunar tables erred by at least 12 minutes. (*Phil. Trans.* 154, Vol. 13, p. 405), suggesting he had then made little progress with the problem.

Flamsteed gave a worked example in his collation of the posthumous works of Horrox (Horrox 1673, p. 494) which had, I found, an error of 13 arcminutes, which is similar to that in the above *DOS* example.

VIII. The First Computation

The first ever *TMM*-based calculation on record was published in the *Philosophical Transactions* of 1710, by "the Reverend Mr H. Cressner, M.A., Fellow of the Royal Society." He also gave a computation based on *DOS*, as published by William Whiston. The occasion was a lunar eclipse observed at Streatham in South London. Thus, at the dawn of the Age of Enlightenment, there existed two rival British lunar theories. Mr Cressner made the claim that he was the first to do this:

> There being therefore no Examples of any Calculation (that I know of) according to that Theory, nor of the Theory's Agreement with Observations yet made Public; I thought it proper to offer this one to this learned Society's perusal...I have added the Calculation from the famous Mr *Flamsteed's* Tables, according to *Horrox's* Theory, as I find them published in the Ingenious Mr *Whiston's* Astronomical Lectures, with the *Radix's* of the Mean Motions, corrected according to their first Author's later Observations, which are the same as Sir *Isaac Newton's* Theory.
>
> By comparing these two Calculations we may observe, that tho' most of the additional Equations in Sir *Isaac Newton's* Theory be very small in this situation of the Moon, yet they all conspire so as to make its Place considerable more agreeable to Observation, than those of *Horrox's* System. (Cressner 1710, p. 16)

It is here implied that *TMM*'s mean motions derived from Flamsteed, as is likely enough. We also gather that Flamsteed had not developed his lunar theory beyond what he published in 1681, except for such slight modification of his mean motions.

We are told that both computations start from the same mean motions. The TMM-PC and DOS-PC programs were here used, the latter with *TMM's* mean motions, for the time given. The two calculations as shown by Cressner purport to ascertain the beginning and end of the eclipse. To my knowledge, *TMM* cannot be used for such, but only for the moment of exactitude. We take what Cressner called "The Mean Time of the True Opposition" for the Julian date of 2nd February 1710 10 hrs (i.e., 10 pm), 54 min, 48 sec at Streatham.

The steps of equation agreed tolerably well, and show that Cressner adopted the correct sign for the sixth equation. His finally-equated value (M_7) differed from TMM-PC by 48 arcseconds, while for the *DOS* program it differed by a mere 25 arcseconds. The latter ought to be smaller, since tables then existed for the *DOS* procedure, while none then did for *TMM's* procedure. We cite the final values for M_7:

Longitudes for Feb. 2nd 1710, 10 hrs 54 mins 48 sec GMT:

TMM-PC	145° 1′ 12″
TMM (Cressner)	145° 00′ 24″
DOS-PC	144° 55′ 51″
DOS (Cressner)	144° 55′ 23″
Actual	*145° 2′ 9″*
Eclipse midpoint	144° 45′ (at 10.45 pm.)

The errors for TMM-PC and DOS-PC respectively amount to $\frac{1}{2}'$ and 5′. Cressner concluded, with regard to the *ending* of the eclipse:

> The Error therefore of Sir *Isaac Newton's* Theory is by this Observation but half a Minute, or none; of *Horrox's System,* Nine Minutes and a half. (Cressner 1710, p. 19)

For the time specified, a few minutes after eclipse exactitude, for which Cressner presented his computation, Flamsteed's method erred by five arcminutes.

IX. Some Conclusions

We can now resolve certain issues that have remained conjectural for almost three centuries.

1) Newton's 1702 lunar theory had an error of -0.5 ± 3.8 arcminutes in the

1680s, taking mean motion values as corrected in 1705. Its mean error gradually increased with time.

2) This was almost twice as accurate as the lunar theory published by Britain's expert on the Horrox theory, Flamsteed, two decades earlier.

3) Halley's adjustment of the Horrocksian eccentricity function was the most important single improvement in Newton's construction of *TMM*, decreasing its mean error by nearly sixty percent.

4) *TMM*'s four new equations were all sound, except that the sixth had its sign the wrong way round, and their coefficients were close to optimal, in giving agreement of the method of computation with observation.

5) Newton went as far as anyone then could have done in discerning such ancillary equations as were capable of improving the Horrocks theory.

6) With the sixth equation adjusted as specified by the 1713 *Principia*, *TMM*'s standard deviation was no more than a mere 1.9 arcminutes (or, a confidence limit of ±3.8 arcminutes).

7) The next chapter will show that *TMM*'s accuracy increased at syzygy positions, traditionally the most important.

Historically, Flamsteed's assessment of *TMM*'s longitude accuracy in the years prior to 1713 was sound while Halley's was, over that period, mistaken. It was not within the bounds claimed by Gregory or Halley. Furthermore, Newton himself misjudged the matter, as evidenced by the several public statements of his on the subject discussed in Chapter One; we may also note an ascerbic recollection by Flamsteed, of a meeting at Greenwich on April 12th, 1704. At first he and Newton disagreed on optical matters (relations being less than cordial), after which:

> I showed him also my new lunar numbers, fitted to his corrections; and how much they erred: at which he seemed surprised, and said "It could not be." But, when he found that the errors of the tables were in observations made in 1675, 1676, and 1677, he laid hold on the time, and confessed he had not looked so far back: whereas, if his deductions from the laws of gravitation were just, they would apply equally in all times."(Letter to Abraham Sharp, Baily 1837, p. 217)

In fairness, however, Newton had avoided making any claims about gravitation in the context of *TMM*, though David Gregory had averred that such a link existed in its Foreword. This report therefore appears as an early expression of what became a widespread viewpoint, albeit made with some scepticism. One can only regret the disappearance of the papers which Flamsteed showed to Newton on this occasion.

Chapter 11

Error Patterns

I. The Error Envelope

To construct error-graphs using the *TMM* program, one runs it repetitively for a given time-increment, thereby creating a table of sequential, computer-derived lunar longitudes. A "macro" was written to accomplish this, which moves the time-value on by a fixed increment after printing out the corresponding longitude value. To start with, the program was set to subtract mean lunar motion from the finally-equated position, at daily intervals. This gave the ellipse function shown in Figure 11.1, i.e., the lunar equation of center, whereby "true anomaly" moves nearly seven degrees away from the mean position. It oscillates to the anomalistic month period.

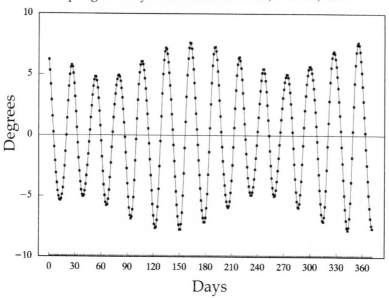

The Ellipse Function
Sampling at Daily Intervals from Noon, Dec. 31, 1680

Figure 11.1: Mean lunar longitude subtracted from *TMM*-generated values, for 365 days.

Over a period of a year, the figure shows how the maximum values of the equation of centre vary. Its envelope has its amplitude vary with the "Horroxian year" cycle, twice per thirteen months.

To construct the error-envelope of *TMM*, a lunar longitude program accurate to arcseconds in historical time was used, able to generate columns of positions for a specified time-interval. The program was set for the identical times as employed in the *TMM*-iteration procedure. Its columns of longitude data were then placed adjacent to those generated by *TMM*, and the error values {*TMM* - modern} were found by subtraction.

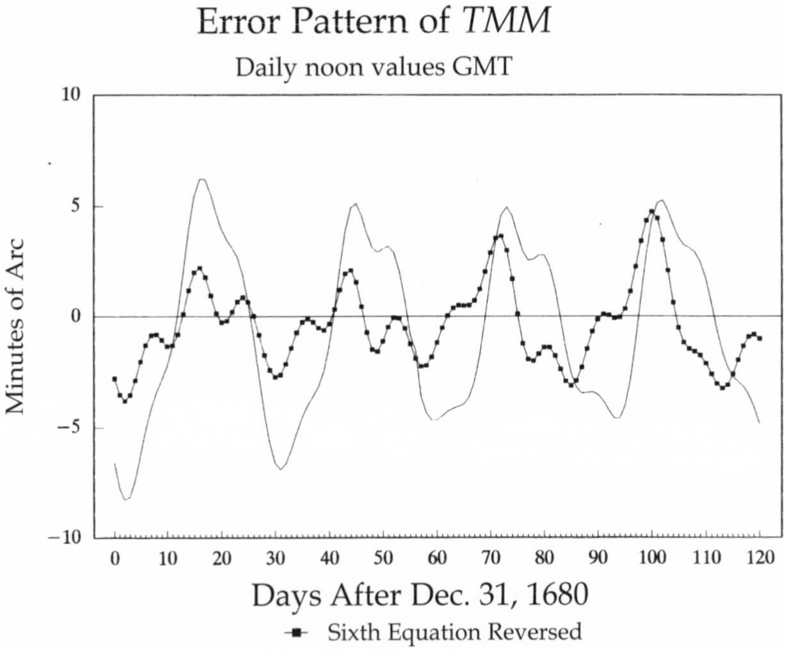

Error Pattern of *TMM*

Daily noon values GMT

Days After Dec. 31, 1680

—■— Sixth Equation Reversed

Figure 11.2a: Comparison of daily error-values of original 1702 version of *TMM* (thin line) with that obtained after reversing the sixth equation (dotted line).

Figure 11.2a shows the graph of the two versions of *TMM* that were discussed in the previous chapter: the original of 1702, and that same program adjusted by reversing the sixth equation and slightly increasing its amplitude, as specified in 1713. A lunar-monthly rhythm is apparent, here peaking at the first lunar quarter, though this is not a permanent feature. A 50% decrease in mean error has appeared from reversing the sixth equation.

Figure 11.2b shows the pattern continued somewhat longer, over eight lunar months.

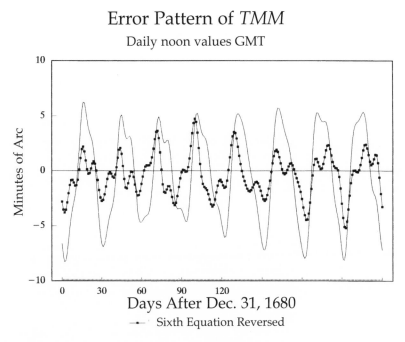

Error Pattern of *TMM*
Daily noon values GMT

Days After Dec. 31, 1680

–•– Sixth Equation Reversed

Figure 11.2b: As before, but sampling over 240 days.

II. Error-Periods

What error pattern is generated by sampling periodically at apogee positions? *TMM* has several terms of the form $(M - A)$ and $(S - A)$, so one should expect rhythms to be discernible at these periodicities. The patterns generated are shown in Figures 11.3a and 11.3b.

We start by locating a time of lunar conjunction with the apse line as the zero position. Mean positions are used, as they are required to stay in position over a period of time. Increments of 27.66 days were added, proceeding through a complete apse revolution of nine years. A large-amplitude rhythm of about six arcminute amplitude appears, undergoing seven cycles per apse revolution, of period 460 days. The astronomical motion generating such a cycle, *TMM*'s strongest periodicity, is obscure!

TMM's Apogee Error
Over a 9-year Apse Cycle

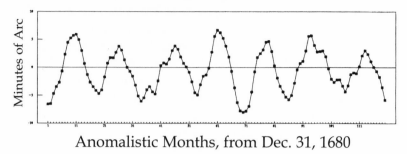

Anomalistic Months, from Dec. 31, 1680

Figure 11.3a: Monthly *TMM* errors sampled at each mean lunar apogee, over a nine-year apse cycle.

TMM's Sun/Apse Error
Sampling every 206 days

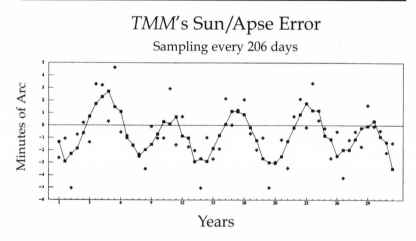

Years

Figure 11.3b: *TMM* errors on successive Sun/apse conjunctions every 6.5 months over a thirty-year period. A three-point moving average has been added.

Sampling instead at conjunctions of the mean Sun and apse, that is every $6\frac{1}{2}$ months or 206 days—another period strongly encoded into the *TMM* program—then the rhythm shown in Figure 11.3b appears. The graph shown presents a six-yearly rhythm. A three-point moving average has been put through the data to smooth it. The effect is weaker than the previous monthly-iteration cycle, having a smaller amplitude of merely three arcminutes.

If, instead, sampling is done at each Full Moon, a more random pattern emerges. The syzygy errors are smaller than usual, as shown in Figure 11.3c, being mostly within two or three minutes of arc: Halley's claim made about the accuracy of *TMM* in his afterword to the third edition of Streete's *Astronomia Carolina* (Streete 1716, p. 70) here appears as valid. And yet, the Full Moons do have an error-rhythm, albeit quite a weak one. Sampling was here done on alternate Full Moons, over a nine-year period, and a five-point moving average put through the data. This time (with some relief) we are able to identify the error-rhythm, as that of the nine-year apse cycle.

TMM's Full Moon Error
Sampling every 59.06 days

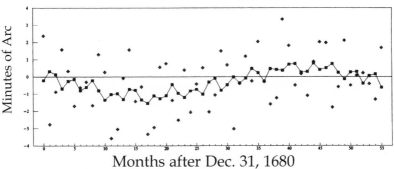

Months after Dec. 31, 1680

Figure 11.3c: *TMM* errors on alternate Full Moons over nine years, plus five-point moving average.

The differing amplitudes of these error-rhythms confirm that *TMM* is primarily linked to the anomalistic cycle, via its functions involving $(A - M)$ and $(A - S)$, and contains little by way of synodic terms, involving $(S - M)$. The anomalistic month error pattern had an amplitude of up to seven arcminutes, which is two or three times the claimed maximal error of the system. Furthermore, it is a coherent rhythm, in contrast with the other two which required smoothing with moving averages to discern them.

III. The "Hidden Terms" of Longitude

It was found in the previous chapter that the modern equations for lunar longitude had four terms in the 2–3 arcminutes range, not included in *TMM*, i.e., that there are modern longitude terms within *TMM*'s amplitude range, not evidently incorporated into it. These did not seem, however, to be related to its error-pattern. The terms all have different periods, and so align now and then, giving an amplitude of up to ten arcminutes, while the *TMM*-2 function has only half such a maximal error. A puzzle thereby arises.

The *TMM* equations contain amplitude-modulating functions. Those for equations two and five vary through the course of a year, while that of the seventh varies through one apse cycle. As the modern equations do not have such, would the comparison be improved by their removal? To find out, TMM-PC2 had its modulating functions removed. This meant that, in the case of the second equation for example, in place of

$$6.25 - 0.31 \cos (H - S_1)$$

merely 6.25 remained. Forty values were generated at 160-day intervals as in the previous chapter, with these modulating terms removed, and compared with correct longitudes. The mean error thereby generated was 0.4±2.1 arcminutes. This is a larger value than TMM-PC2 gives otherwise. From this we conclude, that the three modulating functions within *TMM*'s equations did serve a useful purpose.

To ascertain how many of the modern terms give a comparable accuracy to *TMM*, the first dozen or so of them (see Ch. 10, IV) were written onto the spreadsheet, recalling that their anomaly terms differ by 180° from those of *TMM*. These modern terms are standard, and used in lunar longitude computer programs. Using *TMM*'s mean motions, the four modern variables as above defined were constructed: the anomalies $(M - A + 180)$ and $(S - H + 180)$, the elongation $(M - S)$ and lunar distance from node $(M - N)$. The same error-estimation procedure was used as in the previous chapter, taking thirty times over a 160-day iteration interval, starting with the epoch date of December 31, 1680. The results showed as expected a diminishing error with the addition of successive terms:

The first twelve terms only gave a mean error of −0.44 ±2.1 arcminutes
    "   thirteen   "   −0.48 ± 1.60   "
    "   fourteen   "   <u>−0.46 ± 1.56</u> "
       TMM-PC2   "   −0.38 ± 1.88 " .

It is evident that *TMM* had an accuracy almost equivalent to the first thirteen of the modern equations.

TMM's construction was quite different from the modern set of functions and so there is a limit to how far they can be compared. In the sequence of modern terms, those without any evident equivalence are numbers 8, 9, 10 and 13. It is accepted that the Horrox function incorporates equations 1 and 2 (elliptic function and evection), and presumably also 4. I believe that *TMM* cannot be improved by adding on these equations as extras, though admittedly, equations 8, 9, 10 and 13 have only been so checked altogether and not individually. Around the summer of 1695, when Flamsteed was puzzled that his letters were no longer being answered, Newton had discerned all of the ancillary equations that could have then been found, to the level of accuracy at which he was working.

IV. Latitude

A major criticism of *TMM* by Flamsteed was that it lacked accuracy in latitude:

> The errors in latitude are frequently 2,3 or 4 minutes, which is intolerable. They result not only from my own observations, but from those of others at the same time. (Letter to Caswell, March 1703, Baily 1837, p. 213)

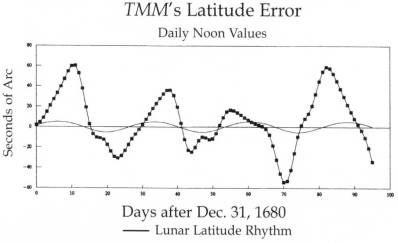

Figure 11.4: Daily errors in *TMM* latitude formula, in arcseconds, with lunar latitude shown for comparison.

To test the latitude formula (Ch. 9, VII), derived from the *TMM* instructions, forty latitude values were derived from *TMM* at 160-day intervals, and subtracted from latitude values generated by the RGO program for those times.

These forty values gave a mean of:

$$\text{latitude error} = -2 \pm 36 \text{ arcseconds,}$$

indicating a confidence limit of within an arcminute. The graph (Figure 11.4) depicts the error of this function, for the opening months of 1681. The error here remains largely within an arcminute. This appears as the second major issue where Flamsteed's judgement over *TMM* has turned out to be erroneous: he was as we have seen also mistaken in his dismissal of the new equations.

Annus Men die	Tempus Appar d h ′	☽ A Rect ° ′ ″	☽ dist a P ° ′ ″	Longitud s ° ′ ″	Latitudo ° ′ ″	diff: a Tab Flam Longit ′ ″	Latit ′ ″	
1692 Maij 16	8.59.11	199.39.20	103.22.50	♎ 23.10.58	4.41.40 A	+0.56	−2.50	
17	9.52.05	213.56.50	108.55.50	♏ 8.01.45	4.59.22 A	+2.13	−2.16	
19	11.46.15	244.36.00	116.04.40	♐ 7.15.47	4.34.50 A	+2.07	−1.24	
Junij 13	7.41.09	208.56.00	107.17.40	♏ 2.58.28	5.04.58 A	+2.06	−2.20	
15	9.30.42	238.28.30	115.15.15	♐ 1.40.05	4.49.25 A	+8.18	−2.18	
16	10.28.37	254.00.00	116.56.20	♐ 15.44.06	4.14.45 A	+7.50	−2.04	
1694 Dec 28	17.30.36	192.18.40	100.34.55	♎ 15.26.18	4.52.15 A	−4.31	−2.24	
30	19.12.57	220.08.30	109.14.30	♏ 14.31.38	3.25.25 A	−3.40	−1.19	
31	20.10.47	235.42.40	112.05.05	♏ 28.29.53	2.16.48 A	−6.02	−2.09	:: ob diei lucem
1695 Jan. 9	3.44.49	358.43.00	84.53.50	♈ 0.51.43	5.11.25 B	+2.11	+1.46	
11	5.18.09	24.12.40	75.13.30	♈ 27.49.40	4.21.13 B	+7.13	+1.01	
12	6.05.04	37.02.40	71.31.00	♉ 10.40.02	3.37.03 B	+8.02	+1.21	
13	6.52.55	50.04.50	68.47.20	♉ 23.12.33	2.41.23 B	+8.31	−0.17	
14	7.41.26	63.18.00	67.04.30	♊ 5.32.35	1·41·03 B	+8.13	+0.08	
18	10.54.32	115.51.00	71.17.00	♋ 24.25.25	2.35.53 B	+7.17	+0.42	
omissa inter- ponatur 1695 Jan. 8	2.57.07	345.42.00	90.31.50	♓ 16.38.05	5.09.34 B	−0.03	+0.44	

differentiæ ostendunt quantum lunæ longitudines et latitudines observatæ supputatas e meis tabulis excedunt vel ab iis deficiunt

J.F:

Altitudo Poli Grenovici 51°.29′.[10]

Table 11.1: Flamsteed's lunar position computations, with dates in Old Style and "Apparent time" measured from noon, i.e., from solar meridian transit; lunar right ascension to the nearest sixth of an arcminute, North polar distance (90° − dec.) converted to longitude (measured 0 − 30°, together with zodiac sign) and latitude marked A or B to distinguish north or south of the ecliptic; plus errors from using his *DOS* tables; as sent to Newton (February 1695, *Corr.* IV p. 85). A transcription error is present: for Dec 30, 1694 longitude should read 13° 41′ instead of 14° 31′.

V. Flamsteed's View of DOS Errors

We may compare Flamsteed's comments upon supposed *TMM* latitude errors with those of his own latitude predictions, as he sent them to Newton in his letter of February 7th, 1695 (Table 11.1). The last column gives latitude errors of his own *DOS* procedure, and their mean error amounts to 1'.4 ± 1'.0. It is odd that he should have regarded errors in someone else's theory of two to three arcminutes as "intolerable" when his own were of the much the same magnitude.

On September 1st 1694, Newton and David Gregory visited the Greenwich Observatory. Gregory's diary recalled how they were shown "about fifty positions of the Moon reduced to a synopsis... Flamsteed is about to show him another hundred," while Flamsteed recalled that the data included "the places of ye Moon derived to ye same times & the differences or errors in 3 large sheets of paper in order to correct the Theorys of her motions." ("Memorandum by David Gregory," *Corr.* IV pp. 7–8). Table 11.1 was presumably a part of that set of data. Sadly, it is all that survives of the lunar meridian readings which Flamsteed sent to Newton during their collaboration from September 1694 to June 1696.

The Table gives clock-time measurements ("Tempus Apparent") of meridian lunar transits, i.e., transit times for when the lunar *limb* first reached the meridian position. An Equation of Time was first applied to give mean time, then, using the notion of the sidereal day, the column of Right Ascension was obtained. "Distance from Vertex," i.e., {90° − Declination} as measured was converted into "Distance from Pole" using a value for the latitude of Greenwich taken as 51° 29'.

The data was converted from topocentric into geocentric form by applying parallax (and refraction) corrections to the vertical "Distance from Pole" reading. Then, using a value for the obliquity of the ecliptic (taken as 29° 30'), Flamsteed has derived longitude and latitude. The table gives longitudes for lunar *centre*, requiring a further correction based on lunar distance.

Thus, the data had to be considerably processed before it was usable to check a lunar theory. Newton once complained to Flamsteed, "I want not your computations but your observations only (June 29, 1685, *Corr.* IV p. 133)." There is no raw data in this Table. Flamsteed's observed data for a meridian transit would have consisted of: inaccurate clock times, up to half an hour out, a solar noon transit for estimating clock error, a vertical mural arc angle, plus instrument correction(s) for that vertical reading.

Table 11.2: Analysis of the Flamsteed Longitude Data (Feb. 1695). Columns show: dates, "GMT" in hours, minutes and seconds reconstructed from Flamsteed's LAT column in Table 11.1, longitudes from ILE program and by Flamsteed from his tables (minutes and seconds only), and three difference columns showing: accuracy of observational data, historic estimate of theoretical errors, and reconstructed errors in the theory.

Date	G.M.T.	ILE Long.	F(DOS)	ΔF(obs)	ΔF(DOS)	Δ(DOS-PC)
1) May 16, 1692	20/55/45	203° 11′ 10″	10′ 02″	0′.2	0′.9	1′.9
2) May 17, 1692	21/48/46	218° 1′ 19″	59′ 32″	−0′.4	2′.2	2′.3
3) May 19, 1692	23/43/12	247° 15′ 42″	13′ 40″	−0′.1	2′.1	1′.2
4) Jun 13, 1692	19/42/54	212° 58′ 23″	56′ 22″	−0′.1	2′.1	6′.6
5) Jun 15, 1692	21/32/53	241° 40′ 39″	31 ′47″	0′.6	8′.3	8′.3
6) Jun 16, 1692	22/31/01	255° 44′ 49″	36′ 16″	0′.7	7′.8	8′.0
7) Dec 29, 1694	5/38/07	195° 28′ 03″	30′ 49″	1′.7	−4′.5	−1′.3
8) Dec 31,1694	7/21/19	223° 42′ 12″	45′ 18″	0′.6	−3′.7	−3′.2
9) Jan 1, 1695	8/19/34	238° 31′ 35″	35′ 55″	1′.7	−6′.0	−3′.9
10) Jan 8, 1695	15/08/27	346° 36′ 9″	38′ 02″	−1′.9	−0′.1	−1′.3
11) Jan 9, 1695	15/56/28	0° 50′ 21″	49′ 32″	−1′.4	2′.2	0′.7
12) Jan 11, 1695	17/30/22	27° 48′ 15″	42 ′27″	−1′.4	7′.2	5′.3
13) Jan 12, 1695	18/17/33	40° 39′ 25″	32′ 00″	−0′.6	8′.0	7′.1
14) Jan 13, 1695	19/05/40	53° 12′ 28″	4′02″	−0′.1	8′.5	8′.4
15) Jan 14, 1695	19/54/25	65° 32′ 50″	24′ 22″	0′.3	8′.2	8′.9
16) Jan 18, 1695	23/08/29	114° 23′ 29″	18′ 08″	−1′.9	7′.3	5′.1
				−0′.1±1.1	3′.1±4.8	3′.4±4.2

ΔF(obs)   = ILE   − F(obs)   Our reconstruction of data accuracy
ΔF(DOS)   = F(obs) − F(DOS)   The historic errors seen by Newton & Gregory
ΔF(DOS-PC) = ILE   − PC(DOS) Modern reconstruction of theoretical errors.

Flamsteed preferred not to reset his Tompion pendulum clocks each day, giving his actual clock times in the *Historia Coelestis* Volume II, sometimes with the corrected clock times adjacent (Kollerstrom & Yallop 1995, p. 240).

The "longitude error" column in the Table has a mean value of 3.1 ± 4.8 arcminutes, notably less than the value ascertained in the previous chapter for the *DOS* model, of ± 6.5 arcminutes, as could suggest that positions generating larger errors on his theory had been removed. This column was intended to indicate how well his version of Horrocks's theory could work. The error column displays a large systematic error, as inherent in his mean motion. (The Table shows a contrast between the three early-morning observations, December 28–31, 1694, and the rest, but that is a mere coincidence, as the program has no diurnal component). These are the error-values which Newton and David Gregory were shown in their September 1694 visit to the Greenwich Observatory.

It will be recalled that the *DOS* mean lunar motion was almost three arcminutes less in value than a "true" mean inferred from the modern Meeus equations (Ch. 5, II). The Table 11.1 has a sign convention whereby the "difference from Flamsteed's Tables" columns on the right represent {observed longitude − theoretical longitude}. For example, for June 15th, the longitude was given as 1° 40′ in the sign Sagittarius, while our DOS-PC model gives 1° 32′ of Sagittarius. This excess of eight minutes in his theory was expressed as +8 arcminutes in Flamsteed's table. It is thereby evident that the systematic error of plus three arcminutes in his longitude error column accords with the error in mean motion which in Chapter Five we expressed as almost −3′ for *DOS* in the 1690s.

A breakdown of Flamsteed's table is given in Table 11.2, where the first step was to reconstruct the GMT times from the clock times as given, using a modern equation of time algorithm (Hughes, Yallop & Hohenkerk 1989, p. 1531). The first columns give the dates and the estimated GMT values in hours, minutes and seconds. These are followed by two longitude columns, firstly as generated using the ILE program, and secondly a reconstruction of Flamsteed's *DOS* longitudes (minutes and seconds only) for those times. The latter was obtained from Table 11.1, subtracting his longitude error estimates from his observed longitudes. These two columns are derived from theories: one using the historic three equations and the other, the modern sixteen hundred.

The ΔF(obs.) column has the difference between ILE longitudes and those in the above Flamsteed table. This gives the only estimate that exists of the accuracy of the raw data from which Newton derived his theory, a standard deviation of one arcminute. We may consider to what extent this was good enough, as *TMM's* third equation had an amplitude of less than an arcminute. Curiously, there was virtually no systematic error in the longitudes derived from observation (8 arcseconds). The above Table is regrettably our sole record of Flamsteed performing such computations, putting his data into a form needed by a theoretician, and as such is of especial interest.

The next column $\Delta F(DOS)$ repeats that given in the previous table, giving Flamsteed's perception of the errors in his own theory. It has as was found an unduly low mean error value of 3′±4′.8. By comparing this value with that from the previous column, we may conclude that this longitude data was some five times more accurate than predictions from the best theory available, clearly leaving scope for theoretical improvement.

Lastly, to check our model of Flamsteed's *DOS* program, and also to confirm that the *DOS* procedure was here being used, as one would assume, a PC(*DOS*) longitude value was obtained for each of these dates. Noon times were used for simplicity, as the error-pattern of such a lunar theory does not vary greatly with time of day. The last column shows the errors it generated, as {PC(*DOS*) − ILE} values, these being slightly smaller than those ascertained by Flamsteed from his computations. Our program will tend to be more accurate than historic computations done manually using tables, the latter being subject to errors of interpolation

Chapter 12

TMM *in the* Principia

The seven steps of *TMM* became embodied in the Second Edition of Newton's *Principia*, in the Scholium to Proposition 35 of Book III (*PNPM* pp. 658–664). This Scholium appears midway through the lunar arguments of Book III, following a summary of lunar inequalities in Proposition 22, and three sections deriving the Variation from gravity theory (Propositions 26, 28 and 29), and before the treatment of the Moon's influence upon the tides. The aim of its text was subtly altered, such that the prediction of longitude was no longer its primary goal.

The Scholium began with the affirmation:

By these computations of the lunar motions I was desirous of showing that by the theory of gravity the motions of the moon could be calculated from their physical causes. (*PNPM* p. 658)

The word "theory" has here a different meaning from that used by Gregory in his title of 1702, *Theory of the Moon's Motion*. Whereas the text of 1702 had been prefaced by Gregory's claim that "Physical Causes" had been reached at last, here that claim was made by its author. However, the Scholium apparently retained *TMM*'s function of finding the longitude. The 1713 text served two different but hopefully concurrent purposes. After describing the seven steps, it averred:

Sic habebitur locus verus Lunae in Orbe, & per reductionem loci hujus ad Eclipticam habebitur Longitudino Lunae. (*PNPM* p. 663)

(Thus you have the true place of the Moon in her orbit, and by reduction to its place in the ecliptic will be found its longitude.)

There was no mention of latitude, and indeed the paragraph making this affirmation vanished from the Third Edition.

The 1713 Scholium omitted what had previously been all-important, namely the numbered steps of equation. It lacked instructions for the sequence in which the various "equations" were to be performed on the five zodiacal variables; although it did present the seven steps of equation in a sequence, almost identical with that of 1702. The Variation, treated earlier, was briefly recapitulated in the Scholium. The second node equation of *TMM* was omitted. Was it the case that the 1713 text could only be "worked,", i.e., made to give longitude values, by presupposing the *TMM* operation sequence?

If *TMM* resembled a watch, then what appeared in 1713 was more an account of its gears, with a new gear added, rather than their assembly. Conceptually, the Variation is independent of eccentricity, being a deformation suffered by a circular orbit from the Sun's pull, and as such was presented in the *Principia* as a successful application of the three-body problem to "explain" the inequality discovered by Tycho Brahe. It was therefore treated prior to the other *TMM* stages. A summary of where *TMM*'s seven steps reappeared in the final Third Edition of 1726 may help:

1726	1702 Equivalents
Book III, Prop. 29	Eq. 5 (the Variation)
Prop. 35, Scholium Para 1:	Eq. 1 (lunar annual equation)
Para 2:	Eq. 1 (other annual equations)
Para 3:	Eq. 2
Para 4:	Eq. 3
Para 5:	Eq. 4 (Equation of Centre)
Para 6:	Eq. 4 (Equation of Centre, epicycle)
Para 7:	Eq. 6

There remain seven stages! All but the last of the above seven paragraphs in the Scholium began with a phrase like "By the same theory of gravity…" or "because of the Sun's force…" The kinematics of *TMM* was transformed into a new dynamics, with the *cause* of the equations given, in terms of forces. We can to some extent retrace the steps of the new approach.

I. Cotes' Contribution

The Scholium into which *TMM* metamorphosed in *PNPM* in 1713 comprised ten paragraphs. It had been changed considerably as a result of comments from Roger Cotes, astronomy professor at Trinity College, Cambridge, who assisted Newton in preparing his Second Edition. The correspondence of Newton and Cotes was published by John Edleston in 1850, Edleston contributing some evaluations of Newton's modifications to the theory which remain of interest.

A "New Scholium to Prop. XXXV" (reprinted in *Corr.* V pp. 291–5) was sent to Cotes, probably in the first week of July 1712 in the view of Edleston (1850, p. 109). It comprised twelve paragraphs, of which the first seven opened with the repetitive phrases we have noted, "by the theory of gravity," etc. These presented the first four steps of equation of *TMM*, and added *two* extra epicycles to the fourth: one was a yearly-period epicycle, to be discussed below, while the other was a nine-yearly one (in its seventh paragraph), varying the rate of motion of the apse line in relation to the Earth's aphelion. It omitted any discussion of the last three *TMM* equations.

There is an undated manuscript entitled *Theoria Luna*, published in the *Correspondence* (IV pp. 1–5) probably belonging to this same period (see Chapter 9, section VIII). It discussed only the first three steps of equation of *TMM*, and also a nine-year inequality to the apse motion. Its logic is comparable to that of the "New Scholium," partly because both showed Newton contemplating a long-period epicycle.

Newton and Cotes discussed the yearly epicycle which was being added onto the Horrox-wheel. A letter of 20th July 1712 found Cotes apprehensive as to when the new emendations to the lunar theory would arrive. Finally, a "revised draft" arrived, undated, inserted into the Correspondence's mid-August 1712 period (V, pp. 328-9). This draft curtailed the fifth paragraph, re-cast the sixth, and added a new seventh, containing the sixth equation. The seventh paragraph now began, "Computatio motus hujus difficilis est...," ("Computation of this motion is difficult...,") instead of referring to the theory of gravity.

Sometime after that, an eighth paragraph must have been sent, starting "Si computatio accuratior desideratur..." It alluded to the Variation, which maximised in the octants, and then proposed an adjusted seventh equation, which it called the "Variatio Secunda," which maximised at the quadrants. It appears that Newton found it harder to justify his sixth and seventh equations by reference to his gravity theory, which is why they were missing from the original "New Scholium" draft. Thereby the Variation became the *sixth* equation, preceded by what had been *TMM*'s sixth equation, now the fifth. We summarise this as follows:

The 1713 sequence	*The* Principia *names*	*the* TMM *steps*	
3rd para	Aequatio semestris	2nd	$\sin 2(A - S)$
4th para	Aequatio semestris secunda	3rd	$\sin 2(N - S)$
5th & 6th paras	Aequatio centri	4th	
7th para	Aequatio centri secunda	6th	$\sin (S - M + H - A)$
8th para	Variatio Prima	5th	$\sin 2(S - M)$
8th para	Variatio Secunda	7th	$\sin (S - M)$

II. An Improvement on TMM?

From Edmond Halley onwards, commentators have inclined to the view that the theme of *TMM* was more fully developed in *PNPM* of 1713. In 1732 Halley as the Astronomer Royal wrote:

> The great Sir Isaac Newton had formed his curious Theory of the Moon, a first Sketch of which was inserted by Dr David Gregory in his *Astronomia Physicae & Geometria Elementa*, published at Oxford 1702; and again, in the second Edition of Sir Isaac's *Principia*, which came out in 1713, we have the same revised and amended by himself. (*Phil. Trans.* 37, pp. 190–1.)

In 1977 Craig Waff wrote:

> A revised and much expanded version of the "Theory of the Moon" was published as the new Scholium to Proposition XXXV.... I might further point out (from my own study in progress of the lunar tables based on Newton's "Theory of the Moon") that many table-makers in the early eighteenth century considered the *Principia* version to be more up-to-date (as indeed it was) than the version which Cohen reprints, and consequently used it as a basic foundation for some of the lunar tables which they constructed." (Waff 1977, p. 71)

Waff was criticising Bernard Cohen for supposedly not having appreciated that the Scholium of 1713 was a "revised and much expanded" version of the 1702 opus. On the other hand, William Whewell was a science historian who appreciated the practical significance of *TMM*, and he affirmed that "These calculations were for a long period the basis of new Tables of the moon," referring to *TMM* (Whewell 1857, I, p. 162). His review of these matters did not suggest that the 1713 *Principia* was an improvement, or that it had been utilised as such by astronomers.

Table 12.1 shows the constants of *TMM* as modified in 1713. The lunar Equation of Centre maximal values were omitted from the *Principia*'s Scholium. These would have had to be generated using the Kepler equation from the eccentricities, no simple task. Thereby the *Principia* text provided less of a practical guide to finding longitude than did *TMM*.

Through these arguments, a tropical reference was introduced into the Second Edition of *PNPM*, whereas the First of 1687 had been primarily sidereal. This complicated matters somewhat, as *PNPM*'s quest for "physical causes" had been within sidereal space, this being the inertial reference framework—the immobile sensorium of the Deity, in Newton's language.

Table 12.1: The *TMM* Constants from 1702 to 1725

These constants represent the maximum values of "equations,", i.e., the peak values found in tables. When these values vary, e.g. seasonally, maximum and minimum values are given.

		TMM	PNPM *1713*	PNPM *1725*
Annual Eqns.:	Moon	11′ 49″	11′ 52″	11′ 51″
	Sun	1° 56′ 20″	1°56′26″	same
	Apogee	20′	19′ 52″	19′ 43″
	Node	−9′ 30″	−9′27″	−9′24″
Equation 2		3′ 56″/3′ 34″	same	same
Equation 3		47″	49″/45″	same
Lunar Eq. Centre		7°39′30″/4°57′56″	–	
Equation of Apogee		12° 15′ 04″	12° 18′	same
Ecc^y in 10^6		66782/43319	66777/43323	same
Horrox-wheel size in 10^6		55050 ± 11732	55050 ± 11727	same
Second Epicycle		–	± 352	same
Eq. 5 (Variation)		37′ 25″/33′ 40″	37′ 11″/33′ 14″	same
Equation 6		2′ 10″	−2′ 25″	same
Equation 7		2′ 20″	1–2′	omitted
Mean Motions				
Aphelion in 100 yrs		21′ 40″	18′ 36″	
Moon	1700 epoch	315° 19′ 50″	315° 20′ 00″	15° 21′ 00″
Apogee	1700 epoch	338° 18′ 20″	338° 20′ 00″	same
Sun	1700 epoch	290° 43′ 50″	290° 43′ 40″	same

TMM in contrast functioned within the tropical framework, i.e., the zodiac, as being what astronomers in practice used. Thus, the yearly motion of the nodes is given as 19° 20′ 32″ sidereally in the Scholium to Proposition 33 (*PNPM,* p. 651) while also a tropical-year period is cited for comparison.

III. An Equation of Eccentricity

When Flamsteed originally explained the Horrocks model to Newton in a letter of October 11th 1694, he added: "To make the aequations bigger in winter yn Summer it will be requisite to make the diameter of this libratory Circle bigger in Winter yn Summer" (*Corr. IV,* p. 27). There was a hint that this modulation was Halley's idea, as being mooted between the three of them.

In his letter of November 1st, 1694 (*Corr. IV,* p. 42), Newton agreed:

The excentricity & equation of ye Moons Orbit is sensibly greater in winter then (sic) in summer & seems to be sometimes as great as Mr Halley makes it, but ye law of its increase I am not yet master of, nor can be till I have seen ye course of the Moon as well when her apogee is in ye summer signes as in ye winter ones,

implying that several years of continuous data would be required to ascertain Halley's equation. That inequality was omitted from *TMM*; however, it appeared in *PNPM* of 1713, as its chief innovation (For a discussion of Halley's theory here, which must remain conjectural as he published nothing on the matter, see *Correspondence* V, pp. 296–8, note 3). A new epicycle was added to the *TMM* mechanism, of yearly period, which, placed on the Horrox-wheel, generated a twice-equated eccentricity and apse motion. Cotes drew a helpful diagram, here shown (Figure 12.1), together with the *Principia's* diagram for comparison. The centre of the lunar orbit is now positioned at *F* instead of *D* as formerly.

This yearly expansion and contraction should not be confused with a supposed overall expansion and contraction of the lunar orbit through the seasons, whereby Newton was perceived as successfully having linked the "annual equation" to a "physical cause", i.e., gravity. Rather, it is a perturbation of the orbit that increases in the winter season, at perihelion, with greater eccentricity and oscillation of the apse, from the Horrox-wheel's dilation.

In the diagram, *TC* represents the mean eccentricity, *TD* the eccentricity once-equated and *TF* that twice-equated. The apse "equation" now

Principia figure (p. 661)   Cotes' version (*Corr*.V p. 285)

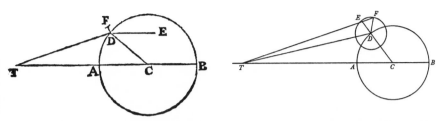

Figure 12.1: The New Epicycle of 1713. The first diagram is from the *Principia*, where F represents the twice-equated position of the lunar orbit centre; the other is Professor Cotes' version, showing more clearly *TF* as the equated eccentricity value.

becomes *FTC*, instead of *DTC*. The wheels have their relative magnitudes specified by: $TC = 5505$, $CD = 1172.7$ and $DF = 35.2$; thereby the lunar eccentricity would fluctuate by ±21% semiannually, while this fluctuation itself varies seasonally by ±3%. Two paragraphs, the sixth and seventh, of *PNPM* explain this new epicycle, its period being given as follows:

> ...and set off the angle *EDF* equal to the excess of the aforesaid annual argument above the distance of the moon's apogee from the sun's perigee forwards; or, which comes to the same thing, take the angle *CDF* equal to the complement of the sun's true anomaly to 360°. (*PNPM* p. 662)

We now show that these two sets of instructions are equivalent. The first can be phrased as:

$$EDF = (S - A) - (A - H + 180)$$
$$= 180 + S + H - 2A$$

$(S - A)$ is the "annual argument," viz. the "Horrox angle," while $(A - H + 180)$ gives the "distance" in zodiac longitude "of the moon's apogee from the sun's perigee." 180° is added, as the *TMM* symbols A and H are measured from apogee and aphelion respectively. The second instruction is given by:

$$CDF = \left[360 - (S_1 - H)\right] \tag{12.1}$$

where S_1 is the "first-equated" solar longitude. However, we know the angle *EDC*, since $DCB = 2(S - A)$,

so $$EDC = DCT = 180 - 2(S - A),$$

and
$$EDF = [360 - (S_1 - H)] - [180 - 2(S - A)]$$
$$= 2(S - A) - (S_1 - H) + 180 \tag{12.2}$$
$$= 180 + S + H - 2A,$$

as above.

The *Principia's* two accounts are identical *only* if we ignore the difference between S_1 and S. This confirms what was said earlier, that *PNPM* was not primarily concerned with the steps of equation. Its phrase "true anomaly" we wrote as $(S_1 - H)$ in equation 2 above. In *TMM* the difference between true and mean anomaly was crucial, here it is not.

This new epicycle responded to the varying distance of the Earth from the Sun, given in the sixth paragraph; I believe that no commentators have remarked upon its function. It amends the equation of centre, so would require a once-equated solar longitude. Cotes was puzzled by this epicycle, and wrote (17 August, 1712, *Corr.* V p. 325):

> It is evident that in the Earth's Aphelium *DF* will coincide with *DC*, & in ye Earth's Perihelium *DF* will coincide with *DH* [*DE* in figure 12.1], so revolving about the centre *D*, that the angle *CDF* may always be equal to the Suns mean Anomaly. Hence the angle *EDF*...will be equal to the excess of ye doubled Annual Argument above the Suns mean anomaly as I observ'd in my last. This is the only way according to which I can apprehend the motion of the point *F* in ye Secondary Epicycle.

Cotes' phrase "the excess of ye doubled annual argument above the Suns mean anomaly" is equivalent to equation 12.2 above—provided we overlook the difference between the sines of mean and true anomaly. At perihelion in midwinter, the Horrox-wheel is required to dilate and be larger than in summer, when *F* is furthest from *C* in figure 12.1 and reaches the point *E* as Cotes observed: in equation 12.1, *CDF* is then 180°.

In the original text of twelve paragraphs another epicycle was added of period nine years, to give eccentricity equated a third time and an apse equated a fourth time. This was subsequently omitted, so we have something to thank Cotes for.

IV. Constructing the New Epicycle

The editor of the *Correspondence* commented as follows upon the new epicycle:

> Though Newton's description does not make this clear, the point *F* is now the empty focus (later centre) of the Moon's

ellipse; it rotates semiannually about the point D. (*Corr.* V
p. 298 note 3)

There are two errors here: the point F cannot be a focus of the ellipse,
because the Earth's centre at T is such, and eccentricity was the distance
between Earth's centre and the orbit centre; nor was the rotation period
semiannual, but was seasonal, conferring a yearly expansion and
contraction upon the Horrox-wheel.

To avoid such confusion, a
further diagram is here presented, in
which A_1 represents apogee and S_1
the sun's position, both once-
equated by their annual equations,
and the three lunar orbit centres,
zero, once and twice equated, are
shown as C, C_1 and C_2.

Figure 12.2: Orbit Centres for *TMM*-
1713.

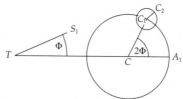

The seventh paragraph of the
Scholium described how to prepare
an extra set of tables to give the angle
FTD, to be added onto the second
apse equation. We shall not proceed
in this manner, but will instead determine the horizontal (i.e., parallel to
TC) and vertical co-ordinates of F
with respect to T (Figure 12.3).
Putting TC equal to 100, CD would
become 21.30 and DF 0.639.
If

Figure 12.3: Reconstructing the
1713 epicycle, in which
$\alpha = CTD$, $\beta = CTF$, $2\Phi = DCB$
and $a = EDF$.

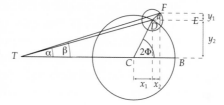

$$y = y_1 + y_2 \text{ and } x = 1 + x_1 + x_2,$$

then

$$y = CD \sin 2\Phi + DF \sin a$$

where $a = EDF = 180 + S + H - 2A$ (equation (2) above) and Φ is the
Horrox angle; and

$$x = 1 + CD \cos 2\Phi + DF \cos a.$$

The eccentricity twice-equated is given by

$$TF = 0.05505\sqrt{(x^2 + y^2)}$$

while the two steps of equation for the apse line are conjointly given by

$$\beta = \arctan \frac{y}{x}$$

For the *TMM* program, we write this as

$$A_3 = A_1 - \beta,$$

replacing the δ function, the angle *CTD*, by β which is *CTF*; likewise the new term for *TF* simply replaces that for *TD*. The point *D* in Figure 12.1 was represented by *F* in *TMM*'s diagram, so the point *F* has acquired a different position, though it retains the same meaning, viz the centre of the lunar orbit.

The eccentricity function has now its own steps of equation, as follows:

$$E_0 = 0.055050 \qquad\qquad = TC$$

$$E_1 = E_0(1 + 0.2130 \sin 2\Phi) = TD$$

$$E_2 = E_0\sqrt{(x_2 + y_2)} \qquad\quad = TF$$

We shall shortly return to this construction, for testing whether any improvement was indeed thereby attained.

V. The Sixth Equation

The seventh paragraph of the Scholium contains the sixth equation, its sign now reversed: "addendam si summa illa fit minor semicirculo, subducendam si major" ("add if their sum comes to less than a semicircle (i.e., <180°), subtract if it comes to more"), the converse of the 1702 instructions. The great error of *TMM* was at last corrected.

On October 31st, 1713, three months after the publication of *PNPM*'s Second Edition, Flamsteed complained to Sharp:

> [Newton's] sixth equation is not allowed by the heavens. He has lately published his *Principia* anew, wherein he makes this equation ablative where it was formerly to be added, and to be added where it was subductive; and has altered his seventh, so as in part to destroy it. (Baily 1837, p. 304).

Flamsteed here erred, as the sixth equation is indeed "allowed by the heavens" once its sign is reversed, as we saw earlier. He returned to this theme in another letter to Sharp of March 20th, 1714, averring that "if I reject them both (i.e., the sixth and seventh equations), the numbers will agree something better with the heavens than if I retain them" (Baily 1837, p. 309).

Baily endorsed Flamsteed's general view as to how modified sixth and seventh equations appeared in *PNPM* of 1713 (Baily 1837, p. 697), as did Whiteside (1975, pp. 323–4) and Cohen (1975, pp. 61–2). In 1989, however, *GHA* stated categorically that "no mention is made of them (i.e., the sixth and seventh equations) in any edition of the *Principia*" (*GHA* p. 267). We cannot endorse the *GHA* view. The sixth equation is clearly specified as having its argument (i.e., angle) formed

> by adding the distance of the moon from the sun to the distance of the moon's apogee from the apogee of the sun (*PNPM* p. 662),

which is the same $(S - M + H - A)$ function about which Whiston had complained in 1703. Its amplitude has increased slightly, by 12% to 2′ 25″, yet it remains the same function.

Admittedly, it was not referred to as the sixth equation, but as the "Aequationum centri Secundam," and described cryptically as "the angle which the line *DF* contains with the line drawn from the point *F* to the moon." (*PNPM* p. 662.) Newton averred, in a letter to Cotes of 12 August 1712 (*Correspondence*, V, p. 320) that

> The Equation described in this Paragraph I had first from observations of Lunar Eclipses, & afterwards found that it answered the Theory of gravity in the manner here described. Its quantity when greatest came to about 2′ 10″ by eclipses. By ye theory tis 2′ 25″.

The suggestion here (whether or not Cotes believed it) is that the 1702 amplitude had been derived empirically, whereas the new amplitude was computed from theory. Lunar eclipses would have given very accurate times at which the $(S - M)$ component was 180°, implying that Newton was thereby able to investigate the nine-year $(H - A)$ apse cycle.

Cohen published two statements upon this matter, before and after Dr Waff's comments, the latter appearing in a book review of Cohen's 1975 publication of *TMM*. In this 1975 opus, Cohen averred, concerning the modification of the sixth equation:

> But after 1713, when Newton had published the above-mentioned correction in ed. 2 of the *Principia*, there was no longer any excuse for continuing to reprint Newton's essay without alteration, as was done in both English editions of Gregory's textbook (and the second Latin edition), and the two English editions of Whiston's *Astronomical Lectures*, even though all declare in their second editions that the text has

been "corrected." Nor was the correction introduced into the later reprintings of the *Miscellanea Curiosa* or of Harris's *Lexicon Technicum*; and it is not even mentioned as an annotation in Horsley's version in his edition of Newton's *Opera*." (Cohen 1975a, p. 62)

It is indeed remarkable, that, as Cohen observed, neither Gregory nor Whiston troubled to introduce this vital correction, in successive editions of their textbooks, without which the function of *TMM* was greatly impaired. These were the main lecturers in astronomy at Britain's two universities, Oxford and Cambridge. Such a fact must serve greatly to endorse Baily's view that *TMM* was not in fact used for some decades after its composition.

In 1980, Cohen merely made the cautious statement that

These results [i.e., *TMM* of 1702] were then corrected and revised and in large measure introduced into the second edition of the *Principia* (1713)." (Cohen 1980, p. 276)

with which one can hardly disagree.

VI. The Seventh Equation

The eighth paragraph of the 1713 Scholium contains a restatement of the seventh equation, as the "Variatio Secunda." The Variation is a sin 2 $(S - L)$ function, while the seventh equation has the form sin $(S - L)$. The 1713 text is:

Ut radius ad sinum versum distantiae Apogaei Lunae a Perigaeo Solis in consequentia, ita angulus quidam *P* ad quartum proportionalem. Et ut radius ad sinum distantiae Lunae a Sole, ita summa hujus quarti proportionalis & anguli cujusdam alterius *Q* ad Variationem Secundam, subducendam si Lunae lumen augetur, addendam si diminuitur. (*PNPM* p. 663).

where *P* and *Q* were assigned magnitudes of 2' and 1' respectively: "As the radius is to the versed sine of the distance *in consequentia* of the apogee of the Moon from the perigee of the Sun, so is a certain angle *P* to a fourth proportional. And as the radius is to the sine of the distance of the Moon from the Sun, so is the sum of this fourth proportional and of a certain other angle *Q* to the second variation, to be subtracted if the light of the Moon is waxing, and to be added if it is waning."

While *PNPM*'s language describing the seventh equation is obscure, Edleston (1850 p. 120) interpreted its meaning as:

$$-\{2'[1 - \cos (A - H + 180)] + 1'\} \sin (M - S).$$

Three sets of brackets within brackets may strain credulity; however, Whiteside in 1975 (p. 325) gave a comparable formula for the modified seventh equation. Its coefficient, i.e., of sin $(M - S)$, he found to be:

$$1' + 2'[1 - \cos (A - H)]$$

There is a 180° shift in the cosine function between the two, equivalent to a reversal of sign. Our *TMM* function was, approximately:

$$M_7 = M_6 + [2' + 1' \cos (H - A)] \sin (S - M).$$

We present the two modern interpretations in comparable form:

Edleston: $\{1' + 2'[1 + \cos (H - A)]\} \sin (S - M)$

Whiteside: $-\{1' + 2'[1 - \cos (H - A)]\} \sin (S - M)$

TMM 1702: $[2' + \quad 1' \cos (H - A)] \sin (S - M)$

From Edleston's formula, sin $(M - S)$ has been changed to $-\sin (S - M)$, and $-\cos (A - H + 180)$ to $+\cos (H - A)$.

Edleston's version of the function had the same signs as *TMM*'s 1702 expression. We adopt his version of the *TMM* seventh equation in the *Principia*, as represents the Newtonian text. We thereby differ from Cohen's view concerning the 1713 version that

this equation (the seventh) is, to all intents and purposes, no longer a part of Newton's system. (Cohen 1975, p. 62).

The language is, however, obscure, and as we have already noted that some astronomers dropped the seventh equation, we should not wholly dismiss Cohen's view. The paragraph containing the seventh equation disappeared from the Third Edition.

Thus all seven steps reappeared in 1713, with amplitudes slightly adjusted, with the sixth and seventh reversed in sequence, with latitude omitted and omitting the second nodal equation. Three components of *TMM* were thus omitted in the 1713 Scholium: the magnitude of the lunar equation of centre, the second nodal equation, which had its own epicycle, as would have affected the "reduction," and any latitude procedure. An extra epicycle was added to the Horrox-wheel, supposedly required by gravity theory.

While absent from the Scholium, a qualitative reference to the node equation was present in the earlier Proposition 22:

But the nodes, on the contrary (by Cor. 2, Prop. LXVI [of Book I]), are quiescent in their syzygies, and go fastest back in their quadratures. (*PNPM* p. 611)

That represents the *TMM* node equation, with syzygy meeting the nodal axis twice-yearly (The term syzygy here applies to the alignment of the two nodes with the Sun-Earth axis). Thus the syntax of *TMM* suffered a dismemberment, serving to support the theory of gravitation.

VII. The Truncated 1726 Version

The Third Edition of *PNPM* contained no explicit affirmation that the Scholium to Proposition 35 of Book III was of practical value. The paragraph containing such, the eighth as we saw in the 1713 edition, was omitted, along with *TMM*'s fifth and seventh equations. We refrain from conjecture as to why that concluding paragraph was deleted. Astronomers who used *TMM* could not readily have taken their directions from the Third Edition.

The Third Edition, the only edition to have been translated into English, concluded its account of the *TMM* equations with the cryptic words:

> And from the moon's place in its orbit thus corrected, its longitude may be found in the syzygies of the luminaries.
> (*PNPM* p. 662 (Motte translation).)

This was followed by some considerations of refraction and mean motion. The literal meaning of the sentence is that longitude may be found at fortnightly intervals. The opacity of its meaning may have encouraged a tendency amongst posterity not to see a working mechanism, viz. *TMM*, buried under its gravity theory. It can however be understood by reference to the earlier 1713 edition, as follows.

The above statement concerning syzygies meant, in the Second Edition, that the theory thus far (i.e., up to the sixth equation) was acccurate in those positions only, whereas, once the last two "variation" equations were added, as was done in the following paragraph, it would become accurate over the whole month. Its omission thus fractured the meaning as originally intended.

VIII. No Baricentre Correction

If linkage with gravity theory was the goal, an equation could have been derived from the Earth's monthly path around the Earth-Moon centre of gravity. Such a displacement would affect the Sun's longitude by something resembling the solar parallax each month. Newton had written to Flamsteed in November of 1694 explaining how motion around a common

Earth-Moon centre of gravity would lead to a monthly solar equation, maximum in the quarters:

> The quantity of this angle I do not yet know certainly. Tis not so great as I thought when I was in London. If you assume it to be 16" or 20" & find that by such an assumption ye greatest errors of ye suns place are diminished you may retain yt quantity, till it shall be determined more exactly. (*Corr.* IV p. 43)

Flamsteed missed the point of the argument, replying that

> The parallactic equation of ye Sun is so small it will scarce be sensible by observation a single vibration of ye pendulum is equall to it. (November 3rd, 1694, *Corr.* IV p. 46)

—confounding diurnal motion with that around the zodiac: the 20" proposed by Newton, as motion in solar longitude, takes eight minutes in time. It was far from Newton's view that such a magnitude could be ignored, but for whatever reason no such solar equation appeared in any version of *TMM*.

TMM in its concluding remarks did specify the value of the "Sun's Horizontal Parallax" as 10", but made no use of it. Later in the century, D'Alembert argued that this baricentre equation would affect the Sun's position by 11"–13" (D'Alembert 1754, p. xvii).

In fact, the equation is smaller, by some 8" (*Corr.* IV p. 44, note 11). Newton had initially overestimated the lunar relative mass by two hundred percent in *PNPM* of 1687 (Kollerstrom 1991, p. 185). In 1713 a baricentre computation appeared in *PNPM* with the lunar mass error reduced to an excess of merely 100% (Wilson 1980, p. 60; Kollerstrom 1985, p. 151). Including such a term would have introduced an error as large as the equation, so its omission was just as well. D'Alembert greatly improved upon Newton's mass-ratio value.

IX. Adding the Epicycle

A century after Kepler had banished epicycles from the heavens with a new, physical astronomy, the second edition of the *Principia* employed two. We here reconstruct them expressing their revolutions trigonometrically. Regrettably, the second of the *Principia's* lunar epicycles led to no improvement in accuracy. We use the interpretation given by Roger Cotes, as a check that our construction is sound.

What has here been called *TMM*-2 had but one modification, namely the sign-reversal of its sixth equation. Here we model two further steps of the

1713 version: the adding of an epicycle, designated as *TMM*-E; and the insertion of the various adjusted constants given in Table 1, plus Edleston's version of the seventh equation. This final step increases the accuracy by 1–2%, a negligible amount. Astronomers of the eighteenth century would have noticed no improvement from so doing.

The working of the two combined epicycles in "*TMM*-E" was checked using the positions of the 1679 perihelion (midwinter) for which $t = -13.3$ days and the next aphelion at $t = 169.2$ days. These positions, for $H - S = 180°$ and $0°$ respectively, give simple triangles (see below, Figure 12.4) whereby the equations of apse and eccentricity can be found: as parts in 10^5, the length of CC_2 ranges between 1208 and 1138 respectively, i.e., 1173 ±35 units of eccentricity, with respect to its mean value CT of 5505).

Figure 12.4: Aphelion/Perihelion Positions for the epicycle, showing greatest and least radius of the 1713 Horrox-wheel, with the varying values for eccentricity and apse equation, at the perihelion (1679) and aphelion (1680) positions.

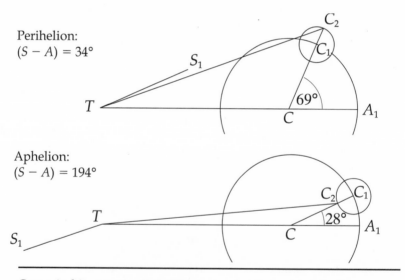

Cotes, in his comments to Newton upon the new epicycle, wrote more than once: "As I apprehend it, the words *additur* and *subducitur* should change places" (*Corr.*p. 285, Cotes to Newton, 3 May 1712). If even the Author experienced confusion on this matter, we should not expect to avoid this ourselves. We start with the relevant celestial longitudes, measuring angles anticlockwise as for the *TMM* diagrams. From simple trigonometry, at these dates of 1679 and 1680:

	Horrox-angle, STA	*Eccy. TC$_2$, (parts in 10^5)*	*Apse equation, ATC$_2$*
Perihelion:	34°	6045	10.7°
Aphelion:	194°	6531	4.7°

The functions f and g in the *TMM* program were readjusted, using the above construction, to give the second-equated apse and eccentricity values. The TMM-2E program took $A - S$ as the Horrox-angle by convention, whereas we have here taken it as $S - A$, making its apse equations negative. Apart from this sign convention, it agreed with the above values. Thus, our function is working as Cotes specified it should.

The graph shown in Figure 12.5 compares error patterns of TMM-2 and TMM-2E, sampling daily over a four month period. It indicates that the latter was less precise. (As before, the program subtracts lunar longitude as given by an accurate modern program in arcminutes from that given by our reconstruction of a historical model). Both versions manifest a synodic error-pattern: adjusting the duration to give four repetitions of the pattern as shown, spanned 118 days of daily sampling, and dividing this period by four gives 29.5 days.

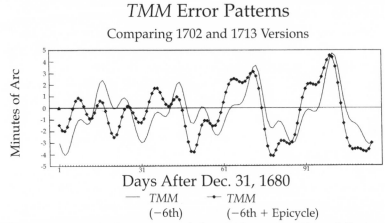

TMM Error Patterns
Comparing 1702 and 1713 Versions

Minutes of Arc

Days After Dec. 31, 1680

— *TMM*   ·—· *TMM*
(−6th)   (−6th + Epicycle)

Figure 12.5: Diminished accuracy of the 1713 version (dotted line), compared with that of TMM-2, over daily sampling January-April 1681.

Error-values were sampled every 160 days, taking this period for the reason given earlier, that it was not near to multiples of the main periods of *TMM*. This gave, for groups of forty:

TMM-2	−0'.4 ±1.9
TMM-2E	−0'.3 ±2.2

The new epicycle was given a reverse direction such that it still aligned with CC_1 at aphelion/perihelion, which gave just as good a result:

$$\text{TMM-2}(-\text{E}) \quad -0'.5 \pm 2.1$$

In other words,resetting the function out of phase made no difference, further evidence that it does not work. We thereby conclude, that the notion of a seasonal expansion and contraction of the Horrox-wheel, Edmond Halley's second contribution to *TMM*, was erroneous. What we have called TMM-2 was the best version for astronomers, who had *nothing to gain* from the greater complexities propounded in the *Principia.*

The notion discussed by the three astronomers back in 1694 was an expansion and contraction of the Horrox-wheel. Putting the radius of that wheel equal to

$$1173 - 35 \cos (S - H) \quad \text{parts in } 10^5$$

would give the required maximum value in winter, and the minimum in summer (when $S - H = 0°$), a somewhat simpler construction. Testing as before every 160 days, still gave no improvement to TMM-2.

At his house in Jermyn Street, behind St James' church in London, the seventy-year old Master of the Mint re-cast his earlier "theory," so that it would more resemble the result of forces interacting between three bodies. Far from being a "much expanded version of the 'Theory of the Moon'" as Craig Waff claimed (1977, p. 71), what appeared in 1713 was a rather abbreviated version. After toying with a nine-year periodicity based upon apse rotation, as two documents probably belonging to this period indicate, he finally decided against it, probably because his reliable data did not extend over a long enough period.

He did however introduce an epicycle, deriving from discussions of the 1690s. Having introduced four valid new equations, and linking up two of the zodiac variables to annual equations, he conferred an annual equation upon the Horrox-wheel itself, in a manner that simply did not work. Thus, what we have earlier called TMM-2 was the optimal format for Newton's lunar theory. We now ascertain to what extent eighteenth-century astronomers, and Edmond Halley in particular, applied these modifications.

Chapter 13

Halley and the Saros Synchrony

Once established as the new Astronomer Royal, Edmond Halley commenced a systematic study of the error-patterns generated by *TMM*. In many ways his study was comparable to what was done in Chapter 11, concerning the *TMM* error-patterns, as will help us to appreciate his viewpoint. It is normally averred that Halley's data from his two decades spent as Astronomer Royal was unpublished and inaccurate. It will here be argued that this data was (a) published and (b) rather accurate. Even more surprisingly, we shall conclude that Halley may have been justified in claiming that his method was accurate enough to win the longitude prize, being the most accurate method for determining longitude proposed anywhere in Europe in the first half of the eighteenth century, though largely ignored by posterity.

The manner in which Halley proceeded is shown in Table 13.1, a page from his 18-year Saros Cycle of observations plus error-estimates, as was published posthumously. Greenwich mean time is specified on the left, together with lunar longitude, plus longitude as predicted by *TMM* for that time, and then the difference between these two is then given in arcminutes. The longitudes are as derived from lunar limb meridian-transit timings, as compared with lunar-centre positions which he computed.

The first line of data, shown here in Halley's table, is for June 21st 1732. On the left is the GMT instant for the lunar limb transit observation; the lunar semidiameter took (we calculate) 1.08 minutes to cross the meridian, adding which would give the time for the lunar centre transit, as was required for computation of its longitude. The true lunar longitude, as reconstructed by a modern program, then was 218° 3′ 50″. This we compare with the value here recorded by Halley, as he reconstructed from his observations, of Scorpio 8° 3′ 36″. His derived-from-observation value was thus within 14 arcseconds, which is pretty good. Consulting our TMM-2 program for his GMT value gives 218° 2′ 52″, indicating that *TMM* was then generating an error of one arcminute. This is more accurate than Halley found, as shown in his right-hand column, and is in the other direction. We return later in the chapter to this discrepancy. What Halley called "Argument. Annum" is the Horrox angle ($S_1 - A_1$). These things give an indication of Halley's accuracy, and of his procedure.

Table 13.1: a page of Halley's error estimates for longitude of lunar centre, for June–September 1732, with GMT given for lunar limb transits, from his *Tabulae Astronomicae*.

LUNÆ MERIDIANÆ LONGITUDINES
GRENOVICI OBSERVATÆ
CUM COMPUTO NOSTRO COLLATÆ.

Anno JULIANO MDCCXXXII. Currente.

Tranſitûs Limbi Lunæ T. æq.				Argument. Annuum.			Diſtantia ☾ à ☉			Longitudo Centri Lunæ Obſervata.			Longitudo Centri Lunæ Comput.			Error Comp	
M.	D.	H.	′ ″	S.	σ	′	S.	σ	′	σ	′	″	σ	′	″	′	″
Junii. 21	7	32	44	9	11	58	3	27	28	♏ 8	3	36	♏ 8	5	8	+1	32
	22	8 21	22	9	12	50	4	9	26	♏20	50	35	♏20	52	6	+1	31
	23	9 13	15	9	13	43	4	21	32	♐ 3	58	20	♐ 3	59	18	+0	58
	24	10 7	57	9	14	36	5	4	0	♐17	29	31	♐17	29	41	+0	10
	25	11 4	22	9	15	29	5	16	47	♑ 1	24	13	♑ 1	24	10	–0	3
Junii. 29	14	46	18	9	19	0	7	10	49	♓ 0	11	19	♓ 0	9	33	–1	46
Julii. 2	17	19	38	9	21	38	8	22	51	♈14	57	41	♈14	54	1	–3	40
	6	20 58	32	9	25	9	10	16	58	♊12	18	24	♊12	13	55	–4	29
	14	2 37	31	10	1	18	1	10	37	♍13	50	26	♍13	50	31	+0	5
::	15	3 18	32	10	2	10	1	21	53	♍26	7	28	♍26	8	20	+0	52
::	17	4 41	54	10	3	55	2	14	32	♎20	32	35	♎20	35	3	+2	28
	20	7 1	59	10	6	33	3	19	47	♏28	4	48	♏28	6	53	+2	5
	23	9 46	0	10	9	12	4	27	44	♑ 8	49	1	♑ 8	47	36	–1	25
	24	10 42	43	10	10	5	5	11	9	♑23	20	20	♑23	19	23	–0	57
	25	11 38	39	10	10	58	5	24	52	♒ 8	17	45	♒ 8	14	44	–3	1
	28	14 21	35	10	13	38	7	7	18	♓24	27	48	♓24	26	26	–1	22
	30	16 7	46	10	15	24	8	5	44	♈24	52	16	♈24	50	1	–2	15
Julii. 31	17	2	20	10	16	17	8	19	42	♉ 9	37	46	♉ 9	34	50	–2	56
Aug. 1	17	58	4	10	17	10	9	3	24	♉24	2	38	♉23	58	55	–3	43
	2	18 54	18	10	18	4	9	16	45	♊ 8	6	56	♊ 8	2	53	–4	3
	3	19 50	5	10	18	57	9	29	44	♊21	52	10	♊21	48	14	–3	56
	5	21 35	57	10	20	43	10	24	34	♋18	31	36	♋18	30	4	–1	32
	18	6 37	1	11	1	20	3	12	27	♐18	41	24	♐18	43	2	+1	38
	19	7 31	27	11	2	14	3	25	9	♑ 2	10	33	♑ 2	10	22	–0	11
	20	8 26	46	11	3	8	4 - 8	15		♑16	9	7	♑16	7	23	–1	44
	22	10 17	11	11	4	55	5	5	37	♒15	41	46	♒15	37	49	–3	57
	25	13 2	6	11	7	36	6	18	41	♈ 2	43	17	♈ 2	39	59	–3	18
	26	13 57	7	11	8	30	7	3	11	♈18	25	45	♈18	22	57	–2	48
	27	14 53	17	11	9	24	7	17	32	♉ 3	50	55	♉ 3	48	37	–2	18
	28	15 50	34	11	10	18	8	1	40	♉18	52	41	♉18	49	55	–2	46
Aug. 29	16	48	15	11	11	12	8	15	27	♊ 3	27	0	♊ 3	23	37	–3	23
Sept. 1	19	33	4	11	13	53	9	24	22	♋14	32	17	♋14	29	25	–2	52
	2	20 22	35	11	14	47	10	6	30	♋27	31	54	♋27	30	13	–1	41
	4	21 53	21	11	16	34	10	29	45	♌22	50	7	♌22	51	23	+1	16

The approach here developed will indicate the benefits of a quantitative study of historic positional astronomy computations, permitting conclusions not drawn from mere opinions handed down. We have just commented on the accuracy of Halley's method in general terms, without evaluating the last column of his tables which shows his estimate of *TMM's* error over the years. Halley's method of applying *TMM* will need to be examined in more detail before attempting that. We here review his application of that cycle to which he gave the name: Saros.

What Halley began, upon becoming Astronomer Royal, comprised the first *TMM*-based computations of a systematic nature. Once he had finalised his procedure, and the tables which it utilised, he was as we shall see not at liberty to adjust it for the following eighteen years. The Second Edition of the *Principia* had made some adjustments to the *TMM* protocol, beyond the mere reversal of the sixth equation, some of which Halley decided to adopt.

The Saros cycle as was known to antiquity was given its name unintentionally by Halley, during his historical researches. He named it in 1691, by mistake (Armitage 1966, p. 126; *Phil. Trans.* 1691, Vol. 16, *Emendationes ac Notae* p. 537; Gingerich 1992, p. 229.). When in 1682 the 26-year-old Halley turned his telescope towards the Moon from his Highbury residence, it was his ambition to follow a complete 18-year Saros cycle, but turbulent events took him to London instead, involving the funding of the *Principia* from his wedding-dowry, which was probably just as well for posterity; however, once established as the Astronomer Royal in 1720 at the age of 64 he recommenced this scheme, that he had first aspired to 38 years earlier.

Halley set forth his notion of tackling the longitude problem in that same article of 1691 (*Phil. Trans.*, Vol 16, p. 536) in which he referred to the Saros. Years later, in 1716, he brought out a third, posthumous edition of Thomas Streete's *Astronomia Carolina*. After printing three years of his sextant observations at the end of the book, he explained his view concerning the Saros cycle (without using that word), whereby it enabled the error-pattern in *TMM*-based tables to be accurately predicted. His reputation for the accurate prediction of eclipses derived, he explained, from this cycle. Cook has described how Halley learnt to predict eclipses using this Saros period (Cook 1998, p. 358). He was now applying his insight into the Saros in a more general manner:

so that whatever Error you found in a former Period, the same is again repeated in a second, under the like Circumstances of the same Distance of the *Moon* from the *Sun* and

Apogaeon...Being thus assured from the Certainty of these Revolutions, that all the intermediate Errors of our *Tables* were not uncertain Wanderings, but regular faults of the *Theories*; I next thought how I might best be inform'd of the Quantity and Places of these Defects.... Nor was there any other way, but from the Heavens themselves, to derive this Correction; by a sedulous and continued series of Observations, to be collated with the *Calculus*, and the Errors noted in an *Abacus*: from whence, at all times under the like situation of the *Sun* and *Moon*, I might take out the Correction to be allow'd. (Streete, Third Edition (1716), p. 68.)

I. The President's Proposal

That was the method. We next hear about it within the pages of the Journal Book of the Royal Society, in May, 1720. Halley had become the new Astronomer Royal, having taken residence in the Observatory two months earlier, and was explaining to the Royal Society the new terms of his employment, for the improvement of the art of finding longitude. Having sailed a ship across the South Atlantic as well as holding the Savilian chair of Geometry at Oxford, he was indeed competent to hold an opinion on the matter. His advice as recorded made no allusion to the Saros concept! His concern was merely for the accurate positioning of zodiacal stars, whereby lunar "appulses" thereto could be used to find longitude.

It was not Halley's view that lunar right ascension and declination could be accurately measured on board a tossing ship: his proposal was that a telescope of up to five feet in length could be used to give accurate measurement of such stellar transits, on a ship. His predecessor had left many gaps in the band of zodiac stars that were necessary for such, he complained, and he proposed to fill these in. Newton's lunar theory (i.e., *TMM*) should be used together with such tables of stellar positions for finding the longitude. He added the fairly evident comment that lunar quarters were optimal for observations because Full Moons were too bright for the timing of stellar appulses. That was all he said.

Sir Isaac Newton, occupying the presidential chair, was moved to comment—and proposed Halley's Saros method! This is a rather curious role reversal. Had we not got Halley's 1716 proposal of his method, it would appear from this altercation as if the whole idea came from the President:

Upon mention made in the above Discussion that it was proposed to use the President's Theory of the Moon's motion for putting the method for the finding of longitude into practice, the President was pleased to observe that he founded his Theory chiefly upon observations of the Moon's place in the conjunctions and oppositions to the Sun, but it would be necessary for the further correction of the Theory, to collect first of all the errors of it in the quadrants, and afterwards what errors there are in the Octants, for which end he proposed it an useful work to frame an Ephemeris of the Moon's motion from the Theory for eighteen years in which period the errors return & this would be a ready means to Examine how much the Theory may Err from the Observations, made at any other time. (Journal Book of the Royal Society, XII, 1720–26, pp. 11–12)

Halley did this. He commenced the vast labour of creating an almost daily ephemeris of lunar positions. After following half of a Saros cycle or one revolution of the lunar apse over nine years, he reported on his conclusions. In the year 1732 when aged 76 he submitted to the Royal Society's journal "A Proposal of a Method for finding Longitude at Sea within a degree, or twenty leagues" (*Phil. Trans.* 1731/2, 37, pp. 185–195). He concluded that his study of the Saros pattern enabled him to improve upon *TMM*, because its errors recurred over the Saros period. He did not suggest that this approach had derived from the Society's President in 1720.

The second Astronomer Royal completed a whole Saros cycle of observations in the year 1739. His posthumously published *Tabulae* contained a section "Precepts for using the Tables," which gave instructions for using his complete Saros of error-computations, by which means, he explained, errors may "in great measure" be corrected. One merely had to find a comparable position in his 18 years, 11 days of observations (or 223 synodic months) as published before or after the date required. Preferably the exact day should be used; however, one could manage with estimating the day before or after.

Halley then went on to claim that, alternatively, one hundred and eleven lunations could be used, as being the period of one apse revolution, roughly half the Saros period, though he admitted this was not so exact. I believe that this method does not in fact work, because no such synchrony then occurs as it does for the Saros, and that this would have undermined the credibility of his high-precision Saros proposal.

No one could investigate Halley's proposal during his lifetime, since he never published his data! Applying his method depended on having almost daily readings such as he was amassing, but no one else had them. He turned out to have a rather similar attitude towards the publishing of his data as his predecessor, though for a different motive. When a stern rebuke was delivered for the neglect of his public duty by Newton from the Presidential chair, at a meeting of March 2nd, 1727, warning Halley that it might be "of ill consequence to continue in the neglect of it," i.e., the presenting of his observations (Baily 1837, p. 188), Halley explained by way of reply:

> he had hitherto kept his observations in his own custody, that
> he might have time to finish the theory he designs to build
> upon them, before others might take the advantage of reaping
> the benefit of his labours.

Having an eye on the longitude prize, he explained, he wished to keep his data until he had perfected the method. His persistence in this attitude for the rest of his life surely goes far towards accounting for the ignoring of his 1731 proposal by posterity as seems to have happened. His observations were not published until 1749, by which time *TMM* had ceased to exercise a formative effect upon astronomers.

Halley had by 1731 amassed fifteen hundred lunar observations over a nine-year period, which is one every two days:

> And that these might be duly applied to rectify the Defects of
> our *Computations*, I have myself compared with the afore-
> mentioned Tables, made according to Sir *Isaac's* Principles, not
> only my own Observations, but also above eight hundred of
> Mr *Flamsteed's*....
> Comparing likewise many of the most accurate of *Mr
> Flamsteed*, made eighteen or thirty-six Years before (that is one
> or two *Periods* before mine) with those of mine which tallied
> with them, I had the satisfaction to find that what I had pro-
> posed in 1710 was fully verified; and that the Errors of the
> *Calculus* in 1690 and 1708, for example, differed insensibly
> from what I found in the like Situation of the Sun and apogee,
> in the Year 1726. The great Agreement of the *Theory* with the
> *Heavens* compensating for the Differences that might otherwise
> arise from the Incommensurability and Excentricity of the
> Motions of the Sun, moon and Apogee. (Halley 1731, p. 194)

Halley nowhere here names the Saros, as neither indeed did Newton in his

1720 comments, only referring to it as the "Period," and *TMM* is referred to familiarly as "the Theory."

In 1735 Charles Leadbetter published the two volumes of his *Compleat System of Astronomy*, then in 1742 his *Uranoscopia*. Halley is mentioned respectfully as the Astronomer Royal, and the virtues of *TMM* are extolled, and the question as to whether anyone has as yet rightly applied it for the preparing of tables is aired, without any mention of Halley's method, and its glossary of terms gave under "Saros" merely a method of predicting eclipses.

II. The Accuracy of Halley's Method

We now evaluate the degree of validity of Halley's claim, using the *TMM* program. The Saros is a period of 223 lunations, or 18 years and ten or eleven days, depending on how many leap-years are involved, plus an extra one-third of a day. It expresses three fundamental synchronies, by what can only be described as a remarkable coincidence, causing the patterns of lunar motion to recur over this cycle:

synodic	$223 \times 29.5306 = 6585.32$ days
nodal	$242 \times 27.2122 = 6585.35$ days
anomalistic	$239 \times 27.5545 = 6585.52$ days.

The sidereal cycle also coincides moderately well, within ten degrees or so (though that is here without relevance), as does the annual cycle since the synchrony falls a mere ten days into the new year, and the apse cycle of just under nine years also coincides fairly well. Halley regarded the latter as quite important, though it has only a very minor function in *TMM*.

A new Moon fell on December 31, 1689, Old Style. We may conveniently start at twelve noon on that date. One Saros cycle takes us to January 11, 1708, 20 hours, and another to January 22nd, 1726, 4a.m. Precision in the time of day is not here required, since *TMM* has no diurnal component to it; however, precision is required in the tie-up between the Julian days on which the *TMM* program runs, and the date used for the longitude program. The dates were checked against the program in the usual manner, using solar longitude to ascertain them correctly.

We thereby model Halley's own investigation, since the above-quoted text cited the Saros-separated years of 1690, 1708 and 1726. 1690 was the date when Flamsteed, with the help of Abraham Sharpe, erected the Greenwich mural arc, recently described by Allan Chapman as being for its time, "the finest and most exact astronomical instrument constructed to-date" (Chapman 1990, p. 57), presumably why Halley chose to start from this date.

Three sets each containing a hundred *TMM* error-values were generated for the test, sampling at two-day intervals, which gave just over six lunar months, separated by 18-year intervals. Modern values of longitude were subtracted from the *TMM* longitudes at each of those 100 times, generating three columns of errors in arcminutes. These had average values of:

-0.6 ± 1.7, -1.3 ± 1.6, -2.0 ± 1.6 arcminutes.

These standard deviations are comparable to those given for TMM-2 in the previous chapter, while the mean values follow the increasing error in *TMM*'s mean motion over the decades (Figure 13.1).

The graph shows these three plotted, greatly supporting Halley's approach. It shows how, over a half-year period, the errors recur exactly according to their position in the Saros cycle. The synchrony of the Saros does indeed provide a key to predicting perturbations, but was it good enough for the longitude prize? To answer that, we next subtract the three error columns one from the other. This was after all Halley's method. This gives three sets of error-differences, which came to:

Saros1 − Saros2	Saros1 − Saros3	Saros2 − Saros3
0′.7 ± 0.32	1′.4 ± 0.64	0′.8 ± 0.32.

TMM and the Saros

Errors in *TMM*–2 over 18-year 11-day intervals

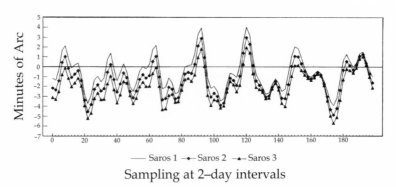

—— Saros 1 —•— Saros 2 —▲— Saros 3

Sampling at 2–day intervals

Figure 13.1: Halley's Saros Synchrony depicting three sets of *TMM* error-patterns, over half-year periods, in identical phases of successive Saros cycles, for the years 1690, 1708 and 1726.

The first of these figures shows a drift of 0.7 arcminutes in mean motion per eighteen years. Apart from this, our use of Halley's method, using one error to estimate another one Saros later, has given a standard deviation of less than one arcminute.

This was by far the most accurate technique of lunar prediction proposed anywhere in Europe in the first half of the eighteenth century. It was a subtle new approach, depending upon periodic return of error-patterns. Whether it was sufficient to claim the longitude prize would depend upon the errors in two sets of observations: one in the present time, and another one Saros earlier. Regrettably, Halley undermined his own case by belittling his predecessor. "A good part" of the merely two minutes of arc error which Halley viewed as *TMM*'s error may have been he felt "the Fault of the Observer." This occurs in the same 1731 report from which we have just quoted. If Flamsteed's observations were so bad, how could readers trust his argument over Saros, which entirely depended on his predecessor's observations? Studies by Yallop and the present writer, indicate that Flamsteed's lunar-limb transit observations were within twenty arcseconds or so (Kollerstrom & Yallop 1995, p. 242).

Part of the error in Halley's method comes from the drift of mean motions. Subtracting this amount out from Halley's error patterns gives the "corrected" graph, showing the marvellous synchrony of the Saros, within

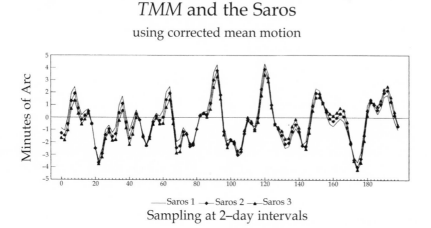

TMM and the Saros
using corrected mean motion

—Saros 1 —◆—Saros 2 —▲—Saros 3
Sampling at 2–day intervals

Figure 13.2: Halley's Saros Synchrony, as before but after subtracting out the error in mean motion over the three successive Saros cycles.

a fraction of an arcminute in its deviation from the *TMM* mechanism! (Figure 13.2.) Subtracting these drift-corrected error columns from each other gave

$$0 \pm 0'.46, \ 0 \pm 0'.39 \ \text{and} \ 0 \pm 0'.86$$

or 29, 23 and 51 arcseconds as standard deviations of their differences.

How accurate were his observations? Halley's lunar meridian transit observations, published in 1749, began in January 1722 and ended in December 1739. He cited G.M.T. values on the left, together with right ascension values for limb transits, over the first five years of observation. Halley then computed right ascension values from *TMM*'s latitude and longitude values for the given times, then took the difference, i. e., the error-value in right ascension.

From December 1725 he changed to a method more convenient for evaluation, giving positions of lunar centre in longitude instead of R.A. His predecessor Flamsteed's observations were all recorded merely as clock or apparent time and not as observed G.M.T., and were for limb transits as observed. Halley omitted declination values, which are not easy to take simultaneously with R.A. for a lunar transit. A copy of his notebook exists at the Royal Astronomical Society's library. His data post-1725 is in the form most convenient for comparison with *TMM*, which was the aim of the exercise.

The longitude accuracy of twenty of Halley's computed longitude positions for the year 1732 I found to be 14" ± 20". Considering that conversions both from apparent to mean time and from limb to lunar centre had been applied, these are plainly the most exact observations recorded up till then within Britain. One is perplexed by the customary comments about inaccuracy and carelessness that historians bestow upon this series of over two thousand lunar transits, with times accurately given in G.M.T. for the first time ever.

As regards the accuracy of his published R.A. values, Bernard Yallop of the RGO analysed the first 35 which Halley published for January 1722, and found their errors to average -10"± 33". They were commonly cited to the nearest arcminute, which may account for their being less accurate than his longitude readings, which began five years later (For further discussion of Halley's method and the accuracy of his observations, see Cook 1996).

Let us summarise Halley's proposed method. It involved two sets of *TMM* computations, and could be used on a day which had a reliable lunar longitude observation of one Saros earlier (or later). The deviation of the old measurement from the *TMM*-computed longitude at that earlier time, conveniently tabulated by Halley, was added on to the *TMM*-computed position for the new position. The method had three sources of error: that in mean motion drift over eighteen years, that in the observations, and that between successive Saros periods in relation to *TMM*. Inherently, the method is accurate to about half of an arcminute, in terms of the third of these errors. This is probably more accurate than Halley himself suspected. 1740 would have been the first year on which his method could be tried using Halley's own data, since his observations started in 1722.

In principle, Halley's method could be used with any lunar theory. If, for example, one removed the four auxiliary equations from *TMM*, then one would merely obtain a larger error-pattern repeating through the Saros cycle, to be subtracted. It was Halley's opinion, however, that *TMM* was the best theory to use for applying his Saros-error correction procedure.

III. The Misunderstanding of Halley's Method

We have argued that science historians have hardly ever recognised the existence of *TMM* as a working mechanism. The problem becomes acute when we seek evaluations of what Halley was doing as Astronomer Royal, as the above-mentioned project then formed his principal occupation. We now quote Francis Baily, who was President of the Royal Astronomical Society in the year 1835, the year in which his *Account* of Flamsteed was published, which did so much to rescue the latter's reputation. The sarcasm of tone is unmistakeable:

> In the year 1731, Dr Halley recalled the attention of the public to an opinion which he had promulgated, about twenty years previously, relative to a proposal for finding the longitude at sea, by means of the motions of the moon: and in a paper inserted in the *Philosophical Transactions* of that year, took occasion to advert to the number of observations of the moon that he had made at the Royal Observatory: which amounted, according to his statement (the accuracy of which I have no reason to suspect), to nearly fifteen hundred. The major part of these observations, however, were made with the transit instrument only: so that declinations remained still to be satisfactorily adjusted. But, it may be amusing to us to know, and may also in some measure lead us to judge of the state of practical astronomy at that day to be informed, that he

considered it a subject of boast and congratulation that, by means of those observations the lunar tables were then rendered so exact that he was "able to compute the true place of the moon with certainty, within the compass of two minutes of her motion, during the present year 1731; and so for the future:" and therefore that *this exactness* was a motive for suggesting it as a means for determining the longitude. The idea, however, was an excellent one: and the method of lunar distances, then in embryo, is now become one of the most important and valuable means of determining the longitude at sea." (Baily 1837, p. 189.)

I suggest that Baily had not apprehended the method that Halley was then proposing. Halley was not claiming that any lunar tables had attained such exactitude, but rather that a method of predicting the errors of those tables could reach such, based on the 18-year Saros cycle of which Baily made no mention.

To suggest that the Astronomer Royal was merely taking right ascension readings, while the "major part" of his declination measurements remained useless because his telescope was not adjusted, implies some degree of incompetence. We merely note that, to compute longitudes as Halley published after 1726 requires both RA and declination readings. To quote from a popular account, "Perhaps because of his age, or because of the equipment used, Halley did not take the great care needed in making the proper adjustments to his equipment." (Heckart 1984, p. 78). In the Armitage biography, Baily's comments are alluded to:

> Baily concluded that no useful purpose could be served by publishing Halley's observations.... Thus the great bulk of Halley's Greenwich observations remain unpublished.
> (Armitage 1960, p. 205)

What Francis Baily said in 1834, in his Presidential Address to the RAS, was that "The astronomical observations, which he [Halley] made in that situation, have never yet been published." (Baily 1837, p. 169) In this he erred: rather, none were *un*published. How did such an idea develop, concerning the over two thousand meridian transit observations of Halley published in 1749, an unprecedented number of unprecedented accuracy?

To substantiate our claim, which may strain credulity, we specify the following. If the clock times as recorded in Halley's notebook, of which a copy exists at the R.A.S. library, are adjusted by applying the Equation of Time (see Howse 1980, p. 38), they will equal the mean times as given in Halley's

Tabulae Astronomicae of 1749. The "Distance a vertice" readings in his note-book (from about 1725 onwards) are two or three degrees from the correct declination readings, implying an instrument correction, possibly specified somewhere in his notes (Zenith distance = 90° − declination). How Baily could have made so awesome an error of judgement, and why successive science historians should have followed him, is not our concern.

The source from which one would expect an authoritative account is Eric Forbes in the Greenwich tercentenary volume (Forbes 1975). Forbes struck a note of scepticism over Halley's method, proposed in 1731:

> This proposal is a repetition of that published in the appendix to the second and third editions of Streete's *Astronomia Carolina*. He [Halley] claims optimistically that the differences between the predictions of the revised lunar theory published in the second edition of Newton's *Principia* (1713) and Flamsteed's lunar observations seldom exceeded ±2'. (Forbes 1975, p. 85)

The phrase, "the revised lunar theory" as published in *PNPM* of 1713 cus-tomarily refers to the inferring of lunar motions from gravity theory. That is the sense in which science historians understand it. The previous chapter evaluated to what extent *PNPM* gave certain modifications to *TMM*, by way of adjusting its parameters, and to what extent it repeated the chain of equa-tions.

Forbes' account gave no hint that eighteen years of observations had been published as the basis for Halley's accurate method of finding longitude:

> Seven years after Halley's death, his *Tabulae Astronomicae* was published in London by John Bevis. These tables, with pre-cepts in both English and Latin, had been submitted to the press by their author as early as 1717 and printed off two years later—before he became Astronomer Royal. In fact, it had been as a result of this appointment that Halley decided to defer their publication so that the lunar tables could be compared with the results of his intended corrections. (Forbes 1975, p. 89)

Forbes' account implied that Halley's tables (as required for computing Halley's version of *TMM*) were printed in 1719, then held back for three decades to allow for their improvement using his new data acquired as Astronomer Royal. The rather important issues here raised will be treated in due course, when we come to the transmission of *TMM*-based lunar theories to France; beginning with Halley handing over certain documents to Delisle on a visit the latter made to London in 1724. Had it been Halley's aim to correct his tables by his long series of observations, then he would at once have noticed the drift in mean lunar motion, from his right-hand column of errors. On the page of positions reproduced above, the mean error in his *TMM* values is −1.4 arcminutes. He did not do so however, as we saw earlier in Chapter Five. Contrary to Forbes' claim, once Halley commenced his immense task of error-comparisons, he would not have wished to readjust his tables any further, since that would have necessitated redoing all the computations.

The French historians D'Alembert and Delambre both described Halley's Saros method, citing reasons as to why it could not work. D'Alembert argued that each of the "arguments on which the inequalities depend" would have to return to the same value at the end of the period, which is plainly not the case, for example the mean anomaly is more than three degrees away while the solar anomaly is more than ten degrees. Halley's method compared residual errors from a theory separated by 223 lunations, as D'Alembert realised: "l'erreur des Tables qu'on en tire doit se trouver la meme dans une seconde période" (D'Alembert 1756, Vol. 3, p. xv), which does not require the above assumption.

D'Alembert had a second criticism, that as the errors recorded by Halley were not generally computed for times exact within 24 hours of the previous Saros (with which they were to be compared), the error may not be "rigorously found." However, D'Alembert should have been able to discern from Halley's tabulations that *TMM*'s error only changed gradually over days. Delambre found similar shortcomings in Halley's method, adding that others including LeMonnier had tried his method without success (LeMonnier 1827, p. 282). I have been unable to find any account of this endeavour by LeMonnier.

Thus the culmination of the Newtonian *TMM* endeavour, to the perfecting of which Britain's most eminent astronomer Edmond Halley dedicated his two decades as Astronomer Royal, so that he could claim to have resolved the most pressing scientific challenge of the age, the finding of longitude—passed into oblivion, remaining to this day unnoticed by historians.

IV. Syzygy Accuracy

Newton's above-quoted belief given in 1720 implied that his theory was most accurate at syzygies and less so at the quarters. The data sent to him by Flamsteed contained no emphasis on the syzygies, but possibly he utilised more the syzygy data. It was traditional for a lunar theory to concentrate on this portion of the month when eclipses occurred. To check this on TMM-PC, fifty successive mean syzygy positions were selected, and fifty square positions, following the epoch date of 1680, and their longitude errors were:

$$\text{syzygy} \quad -0'.21 \pm 1'.52$$
$$\text{quarters} \quad -0'.23 \pm 1'.70$$

That only spans a two-year period, but suggests that its accuracy may indeed have improved around the syzygy positions, as appeared in the error-pattern of Figure 11.3.

V. Halley's Version of TMM

We have previously assumed that Halley used *TMM*-2, as this concurred fairly well with the worked example given in Halley's *Tabulae Astronomicae* (Ch.10, I). Having reviewed the modifications added in 1713, we now re-create more exactly Halley's procedure.

Slight divergences from *TMM*-2 should not affect conclusions reached earlier, since the error-replication characteristic of the Saros synchrony will apply whatever lunar theory is utilised. We now describe what Halley did each day, over eighteen years. Not even the early accounts by Baily (1835, p. 722) or Whewell (1837, II p. 210) recognised this, nor have his biographers appreciated the extent to which the astronomer used Newton's procedure on a more or less daily basis during his tenure of the Greenwich Observatory.

His version of *TMM* is described in his *Tabulae Astronomicae*, under the section "To find the Moon's Place for any given Time" (Halley 1749, p. 59), and is given as a twelve-step procedure. The mean motions are formed from those of *TMM* as indicated in Ch. 5: adding 1' 40" to the mean apse, subtracting 1' from the node and adding 10" to the mean moon, for the 1680 epoch. The first three equations are unaltered, but then *TMM*'s sixth equation was inserted before the equation of centre:

> For the argument of the fourth Equation add the Place of the Sun's Apogee to the Annual Argument, and subtract their sum from the Place of the Moon thrice equated. But this being Sir *Isaac Newton's* sixth Equation,... (Halley 1749, p. 59)

This means that the sine function involved has the "argument" $M_3 - [(S_1 - A_1) + H]$ or $-(S_1 - M_3 + H - A_1)$, the negative of *TMM's* sixth equation term (Ch. 8, I). The next section states:

> In the second column of *Tabula Aequationis Apogai & Excentricatum Orbis Lunae* is the second Equation of the Moon's Apogee; in the fourth, the Excentricity of her orbit; and in the seventh, the Logarithm for finding the Equation of Center, all answering to the Annual Argument. (Halley 1749, p. 59)

The modification to *TMM* given in 1713 is here alluded to, whereby the Horrox-wheel expands and contracts seasonally "answering to the annual argument"; it being evident as we saw from the correspondence of 1694 that this idea came from Halley.

Newton added an extra epicycle to accomplish this, causing the radius of the Horrox-wheel to expand and contract by some 3%, to be largest in winter (perihelion) and smallest in summer (aphelion). His tables have small additional tables to give the small increments according to the $(S - H)$ argument. The previous chapter found no difference in accuracy between these different approaches, both being *less* accurate than the original design without them.

The Variation was his last equation, followed by the Reduction using a twice-equated node. The six steps of Halley's worked example, compared with the computer model of Halley's version, are as follows:

for December 5th 1725 O.S., 9hrs 8m 5s, when t=16410.3806

	Halley's results	*TMM-PC-"Halley"*	*TMM*
M_0	1 21° 25′ 40″	1 21° 25′ 33″	
Eqn 1 (annual)	+2′ 38″	+2′ 36″	
Eqn 2	+1′ 0″	+1′ 2″	
Eqn 3	−0′ 41″	−0′ 40″	
Eqn 4 (− *TMM's* 6th)	−1′ 41″	−1′ 35″	
Eqn 5 (Eq. of Centre)	−5° 3′ 56″	−5° 4′ 5″	−5° 2′ 54″
Eqn 6 (Variation)	−36′ 15″	−36′ 17″	
Reduction	−4′ 11″	−4′ 12″	
M_{end}	1 15° 42′ 34″	1 15° 42′ 21″	
Eccentricity	0.06643	0.06676	0.06643
Apse Eq.	−2° 41′ 0″	−2° 45′ 10″	2° 41′ 2″

The steps of equation tie up tolerably well; however, the eccentricity and apse equation values indicate that Halley has used the simple Horrox-wheel of *TMM*. The month was December, when it would be maximally-expanded if varying seasonally.

Earlier, Halley gave a worked example at a New Moon, treated rather more briefly, with no eccentricity or apse equations specified. We may compare its fifth equation, the "Equation of Centre":

Halley's value	*TMM-2 value*	*TMM-E*
4° 58' 28"	4° 58' 4"	5° 0' 26"

for July 2nd, 1684 O.S., 2h 41m., t = 1279.1119 days. "*TMM-E*" has the 1713 constants plus the extra epicycle. It is again evident that the simple *TMM* procedure has been used. These worked examples indicate that Halley made only two modifications to the *TMM*-2 procedure, as well as modifying its mean motions: he removed the seventh equation and made the sixth *TMM* equation his fourth. This version will be called, *TMM*-H.

A hint of further modifications, probably not of great consequence, appears in a letter by Joseph Crosthwaite to Abraham Sharp, of May 6th, 1720, after the former had been shown a copy of Halley's unpublished lunar tables: "He [Halley] has left out two or three small equations, as Mr Flamsteed has done, and altered the precepts a little for the easier (as he says) obtaining her place." (Baily 1837, p. 335)

We started this chapter with a page of Halley's longitude data. We now recreate the error-patterns of that period to compare with those recorded by Halley. As it was half a century after *TMM*'s composition, we start by checking *TMM*-H's mean motions (Appendix III) for 1700, 1720 and 1740: putting t equal to multiples of 7305 days, as the integer value for twenty Julian years, gives these three sets of mean motion within an arcsecond or so.

The error-values which Halley tabulated with such heroic patience, given in the last column of Table 13.1, were the differences between two computed values: the longitude of lunar centre—computed from meridian-transit lunar limb observations at "time equated," i.e., GMT times as given, using his value of lunar parallax—was subtracted from the *predicted* value of that same parameter, as computed from his tables, using his version of *TMM*.

He did not do this every day, but rather performed it intermittently—2100 observations over one Saros is just under one per three days. His method could still work using such data because, as we have seen, the *TMM* error-pattern has no diurnal component, varying only gradually day by day.

Halley's *TMM*

Daily Error Patterns

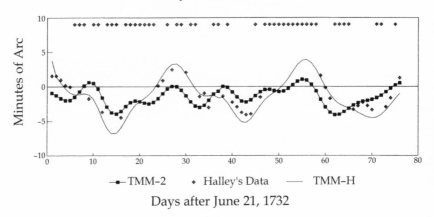

—■—TMM-2 ◆ Halley's Data —— TMM–H

Days after June 21, 1732

Figure 13.3: Accuracy of Halley's version of *TMM*. His own error-estimates, in arcminutes of ecliptic longitude (separate squares), are compared with the computer replicas "*TMM*-2" and "*TMM*-H" (Squares along top of graph represent no-observation days).

Halley's tables were amongst the first to be composed in a manner that was derived from *TMM*, and so the extent to which their values concur with our replica thereof is of interest. Figure 13.3 depicts a typical page of Halley's error-values, showing how they were slightly larger than would have been generated by *TMM*-2 (dotted line), the optimal form for Newton's lunar theory, whereas they are generally within an arcminute of the *TMM*-H error-curve (thin line). Halley's errors deviated by up to an arcminute from our theoretically-constructed error curve, partly due to his observation errors and in part from the interpolation of tables.

In addition, a 1713 version of *TMM* with the variable-radius Horrox-wheel (Ch. 12, IX) was tested for errors over the same range of data as has been plotted in Figure 4a. Standard deviations of the respective error-patterns were as follows.

Halley's page of 34 error-estimates:	±2'.60
TMM-H	±2'.69
TMM-2	±1'.30
TMM-1713	±3'.44

This result establishes what had earlier been surmised, that Halley was not using a variable Horrox-wheel, with its extra equations of eccentricity and apse. Rather, he was proceeding with a six-step method using *TMM*'s sixth equation as his fourth. In 1710 a Mr Cressner became the first person to apply *TMM*-2 (Ch. 10, IX), the optimal format for Newton's lunar theory, reporting this in the *Philosophical Transactions*. Halley did not follow this example, but rather developed his own less accurate version.

Each of Halley's meridian readings (Table 13.1) was taken at a different time of day, whereas the error-curves plotted in Figure 13.3 were for the same time of day (arbitrarily set at 0.4 of a day after noon, or 9.36 pm.).

VI. A Silent Crisis

For most of the Saros period recorded by Halley, errors remain around the two to three arcminutes shown in Figure 13.3. In the spring of 1733 however, Halley's use of Newton's method started to give him errors which were twice that which he had averred before the Royal Society as *TMM*'s maximum. Figure 13.4 plots exactly the same three parameters as before, less than a year after the previous page of data. The 77-year-old Halley was regularly, and faithfully, recording eight arcminutes of error.

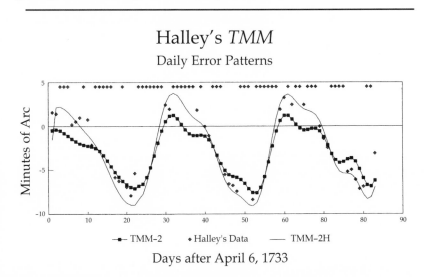

Halley's *TMM*
Daily Error Patterns

Days after April 6, 1733

Figure 13.4: A Silent Crisis. The same variables plotted as in Figure 13.3, for April-June 1733, showing maximal errors arising at 30-day intervals (Squares along top of graph represent no-observation days).

In part this was due to the mean value having accrued two arcminutes of error, so that most of the error-values were negative—a fact analysed years later by D'Alembert, who inferred an error of $-2'$ 10" in Halley's value of mean lunar motion, on the basis of these error-values (D'Alembert 1754, Preface, p. xv). Halley's computations were made every two or three days on average, and the days omitted appear as points along the top of the graph. The graph plots the subtractions performed by Halley, expressed in minutes of arc: his computed lunar centre longitudes derived from observations, and those derived from his version of *TMM*. It is evident that his daily error-estimates concentrated upon the maximal error-periods and the descending parts of the graph, then ceased once the curve began its upward movement. His eighteen years of data tabulation shows no other period of such large errors. One could wish for some comment, but the historical record remains silent.

From these error-values may be derived a further estimate of Halley's accuracy, by subtracting from them the value of *TMM*-H. For the page of Halley's data represented in Figure 13.4, this gave a value of $0'.2 \pm 0'.8$, which may be compared with the above-found mean error in Halley's lunar longitudes as $0'.2 \pm 0'.3$. A small error of about half an arcminute has entered, presumably due to Halley interpolating values from his own tables.

Figure 13.4 shows maximal errors arising around April 26th and on two further occasions after 30-day intervals. They have a 30-day periodicity, coinciding temporarily with the 27.5 day recurrence of perigee. Let us perform what we may hope that Halley did, by comparing the maximal errors in their 30-day periods with those of the Saros before and after. Comparing the noon values then generated by TMM-PC-H, and also at noons nearest to one Saros before and after to these dates, in the usual manner, we obtain:

Maximal Error Values (in arcminutes) at Saros intervals, Compared:

	1715		1733	Δ		1751	Δ
16th April:	−5.8	27th April:	−7.1	1.3	8th May:	−8.2	1.1
+ 30 days:	−6.5	+ 30 days:	−7.4	0.9	+ 30 days:	−8.5	1.1
+ 60 days:	−5.9	+ 60 days:	−6.9	1.0	+ 60 days:	−7.5	0.6
Mean:	−6.1		−7.1	1.1		−7.9	0.9

TMM accumulates a mean motion error of half an arcminute per Saros cycle. What we have called Δ in the above Table, is the *error that would result from using Halley's method*, through comparison with the previous Saros. This reinforces the conclusion of Chapter Twelve, that Halley's method was in principle accurate enough for the longitude prize. His method gives an error of about one arcminute, or half that if the drift in mean motion is removed, even for the maximal error combination in the spring of 1733. (Halley's method could moreover have survived the abandoning of *TMM*: whatever more exact theory was used, its error-pattern should still recur through the synchrony of the Saros.)

Baily claimed that Halley's tables were based on Newtonian rules "as corrected in the Second Edition of the Principia" (Baily 1837, p. 705). Halley occasionally used the adjusted values there given; for example, his tables of eccentricity have their maximal values range from 66777 to 43323, as there specified, and not 66782 to 43319 as in *TMM*. On the other hand, his annual equations used the 1702 values. His method took little of significance from the 1713 version. His tables came to be widely used in France, and it is to France that we turn for further developments.

Chapter 14

Construction of Tables

Newton's proposal to add on half a dozen extra "equations" was highly innovative, doubling the number of tables required, which was why three decades elapsed after *TMM*'s publication before any British tables based upon it were published. The historical record remains scant over the decades after *TMM*'s composition, when the first tables were composed but not published. The key figures appear as Flamsteed, Nicholas Delisle in Paris, and Halley.

In 1712, in his preface to his "Pirate" version of Flamsteed's *Historia Coelestis*, Halley wrote,

> ...the fluctuations of this roving planet [the Moon] doubtless returning into orbit after the cycle of 223 synodic months. Thus the position of the Moon, discovered from the most perfect tables shortly to be published... (Chapman 1982, p. 193)

What were these "most perfect tables"? Regrettably, the present study will not succeed in removing this vital question from the realm of conjecture. The previous chapter showed how the first systematic use of *TMM*-based tables was by Halley in his capacity as Astronomer Royal; of the tables he used and the copies which others made thereof, no trace remains in Britain, all that remains being contained in his posthumously published work of 1749.

I. Flamsteed

A notebook of Flamsteed's in the archives of Cambridge University Library (RGO 1/50H) contains thirty-seven quarto pages of solar and lunar tables, preliminary drafts for which exist in another notebook (1/50 G). Each of *TMM*'s seven steps of equation was there tabulated, with peak values as specified, reaching their maxima at the non-symmetrical positions required. The table for the apse equation, for example, reached its maximum of 12° 15′ 00″ at 51° of the "annual argument." (Ch. 7, III.)

This is the sole surviving set of lunar tables composed in the eighteenth century by Britain's first Astronomer Royal, dated February and August, 1702. Some computations are given in the notebook, followed by an eighteen-step sequence of instructions for using the tables. On the following page is written "July 1714, Burstow," Burstow being Flamsteed's parish church in Surrey. In 1702, some months after *TMM*'s publication, Flamsteed wrote to Abraham Sharp that he had prepared some "new tables" which were "40 quarto pages and upwards." (Dec. 14th, 1702, Baily 1837, p. 210.)

A letter to Sharp of January 1703 (Ch. 5, IV) gave hints about the tables. His procedure

> makes every sign [i.e., 30° interval] of mean anomaly take up a whole page....It cost me and Mr Hudson [Hodgson?] above 3 month's pains to calculate these tables....I have formed the tables for finding the variations with the small inequalities in the same manner.... The tables of the second equations of the apogee and node.... (Baily 1837, pp. 210–11)

A "second equation" of the node can only have come from *TMM*. The allusion to "variations with the small inequalities" must likewise refer to the new equations. His letter to Sharp affirms that his forthcoming *Historia Coelestis Britannica* will contain his newly-constructed tables, so that his pains "will be of great use." This turned out not to be the case.

The letter continued with the strange claim that the tables were based upon his own procedure:

> ...to calculate the moon's place in my correct theory (I call it mine, because it consists of my solar tables and lunar numbers corrected by myself; *and shall own nothing of Mr Newton's labours till he fairly owns what he has from the Observatory;*) and I believe that none but myself would have been at the pains to make so many tables as I have for this purpose. (Baily 1837, p. 210–11)

TMM's "solar equation" derived from Flamsteed (Ch. 6, II), and he evidently believed that he had supplied other of its constants. (letter to Sharp, March 30th 1704, Baily p. 216).

Francis Baily inspected the RGO notebook, after which he wrote a letter to the then Astronomer Royal, Professor Airy. It was written in 1836, the year after the first publication of Baily's *Account of the Revd. John Flamsteed*, and has never been published. It contains a notable shift in his viewpoint and is worth quoting in full:

> My dear Sir,
>
> I herewith return the Ms. of Flamsteed, Vol. 50 I, which contains one of the sets of lunar tables that he composed for his own use. They are founded on the Newtonian Rules given by Dr Gregory in his *Astronomia Elementa*, pages 323–336; and are a curious and interesting historical document, inasmuch as

they are the first that were computed acording to Newton's theory, and afford incontrovertible internal evidence that they are the same as those which, according to Mr Hodgson's account (in his "Theory of Jupiter's Satellites") were *surreptitiously* conveyed into the hands of M. LeMonnier, and published by him in his *Institutions Astronomiques.*

D'Alembert, in his *Recherches sur differens points importans du Systeme du monde*, speaks highly of LeMonnier's lunar tables; & (notwithstanding the then existing tables of Euler, Clairaut & even Mayer) proposes them as the touchstone by which the lunar motions were to be rectified. If any merit however is due, it ought to have been given to Flamsteed. I have a paper on the anvil, in which I shall endeavour to set the public right in this respect.

The tables in pages 51 & 52 were formed 12 years after the preceding ones, & are computed agreeably to the corrected values for the 6th & 7th equation given by Newton in his 2nd edition of the *Principia*. The second of these two tables was afterwards wholly abandoned by Newton in the third edition of the *Principia*. But nothing of these alterations is mentioned by, nor do they appear to have been known to, LeMonnier: who has rigidly followed the first set of tables, which were probably those that had been conveyed to him by some person or persons unknown to us.

Between us yours truly,

Francis Baily

(the letter is folded into the Flamsteed notebook, RGO 1/50H)

Baily never published the paper that he had "on the anvil." How, after all, could he have reconciled his wish to credit Flamsteed for LeMonnier's highly accurate tables, with Flamsteed's dismissive letters averring that *TMM* generated errors of up to 8–10 arcminutes? (to Sharp, 31 Oct 1713, 20 March & 31 Aug 1714; Baily 1837, pp. 309–11). The Flamsteed tables did, however, exert a great influence upon eighteenth-century Newtonian astronomers.

In Baily's *Account*, the tables composed in 1702/3 by Flamsteed were hardly viewed as *TMM*-based (pp. 703–4), and it was therefore not made clear that LeMonnier's procedure was Newtonian. We shall see shortly how the tables of LeMonnier formed the case *par excellence* of *TMM*'s adoption by an astronomer. In consequence, Baily underestimated the extent of

		N.Delisle 1750	*F.Baily* 1835	*W.Whewell* 1837	*C.Waff.* 1977	*G.H.A.* 1989
N. Delisle	1717	*		*	*	
P. Horrebow	1718	*			*	
N. Grammatici	1726	*		*	*	
R. Wright	1732	*	*	*	*	*
A. Capello	1733	*		*	*	
C. Leadbetter	1735	*	*		*	*
R. Dunthorne	1739			*	*	
C. Brent	1740				*	
LiXiang	1742					
P. LeMonnier	1746				*	*
E. Halley	1720/1749	*	*		*	*

Table 14.1: Listings of "Newtonian" astronomers in the early eighteenth century

TMM's influence. Table 14.1 shows how several sources have differently evaluated the astronomers who published textbooks utilising *TMM*. The only major source which Baily perceived was Halley, as Leadbetter abandoned *TMM* in his second publication while Wright confused the procedure somewhat, which explains the rather dismissive tone adopted in the *Account*.

William Whiston, who was on good terms with Flamsteed, affirmed categorically in his Lucasian lectures of 1703 that no tables based upon *TMM* existed (Ch. 1, I), published this view in 1707, and did not alter it in any successive edition. Abraham Sharp's letters of reply to Flamsteed in 1703, and later in 1716 when the construction of lunar tables was again raised, contained no confirmation that he had seen the new tables (Baily 1837, pp. 323, 328).

II. Delisle

The first claim to have composed "Newtonian" tables was made by a Frenchman, Joseph-Nicholas Delisle. The claim was made in the context of the 1713 *Principia*, without reference to the 1702 publication. To quote Schaffer, Delisle was

> committed to the construction of astronomical tables in which
> the celestial mechanics of Newton would be compared direct-
> ly with the best existing observations. (Thrower 1990, p. 265).

Letters of Delisle's from 1717 (unpublished) remarked that Newtonian-based lunar tables could not be found, and that they would be hard to compose because a fresh start was required:

> J'en ai calculé pour le soleil & pour la lune uniquement sur les déterminations que M. Newton a tirées de la théorie de la pesanteur. Mais j'ai trouvé dans la construction de ces tables beaucoup plus de difficultés que l'on n'en a ordinairement dans la construction de pareille tables, lorsque les fondements en sont établis; & cela parce que M. Newton reconnaissant beaucoup plus d'irrégularités dans les mouvements de la lune que les autres Astronomes n'en admettant & attribuent ces irrégularités à des causes physiques, le calcul s'en est trouvé fort long & fort embarrassé.

> (to Teinturier, 7 Feb 1717, Paris Observatory Archives)

Delisle had a prolific correspondence with European astronomers and succeeded Phillipe de la Hire as lecturer at the College Royale in 1718. During this period, "Delisle had already tried and failed to interest the Parisians in Newton's lunar theory as a means for resurrecting mathematical astronomy in Paris." (Greenberg 1984, p. 152.) Delisle was elected as a Fellow of the Royal Society in 1724.

Delisle's tables, in manuscript at the Archives of the Paris Observatory, are entitled merely, "Tables du Soleil & de le Lune suivant la théorie de Mr Newton dans la 2nd Edition de ses Principles." They have no author's name, being classified under "De la Hire," nor do they contain any worked example, making it hard to ascertain how well they worked. His *Lettres sur les Tables Astronomiques de M. Halley* were published in 1749 and 1750, the second of which exists in the library of the Paris Observatory, containing his claim to be the first to prepare such tables (Baily 1837, p. 705). It also gave the list of others as cited in the above Table.

III. Halley

The anonymous Preface to Edmund Halley's posthumous "Tables" published in 1749 averred that they had been "sent to the Press in the year 1717, and printed off in 1719," a view echoed by Baily, that they were "constructed in 1717 and printed in 1719" (Baily 1837, p. 705). For whatever reason, there was a desire to claim that Halley drew up his tables before the death of Flamsteed in 1719. Likewise, Cook has reaffirmed that Halley's tables were "set up in print before 1720." (Cook 1998, p. 366). Contemporary accounts make these dates a little too early, with the lunar tables only completed after Halley became Astronomer Royal in the spring of 1720, i.e., following the death of Flamsteed.

In 1718, an account of lunar tables owned by Edmond Halley reached Delisle. Writing from Amsterdam after a visit to London, the Königsberg astronomy professor G. H. Rast informed Delisle that:

> ...existit a [Halley] quidem perpetua motuum lunarium ephemeris (quae lunae ac solem situs post 18 annos et horas paucissimas, eosdem recurrere praecipue ostendere voluit).
> (1 July 1718; Paris Observatory Archives)

Delisle replied by way of confirmation that Louville

> m'avoit communiqué à son retour d'Angleterre ce que vous appelez l'Ephémeride perpétuelle des mouvements de la lune,
> (Paris Observatory Archives)

and added that Halley was claiming an accuracy of half an hour in past time (equivalent to 12 arcminutes, i.e., no improvement on earlier tables). Decades earlier, Halley had exhibited before the Royal Society such "perpetual tables" for working his Saros method of eclipse prediction (2nd November 1692, McPike 1932, p. 230), so there was nothing new in what was then reported to Delisle.

Halley was proposing to add on various smaller tables, DeLisle added in the same letter, "le tout suivant la theorie de M. Newton." These were expected to limit the errors to no more than 4' (Delisle to Rast 16 July 1718, Paris Observatory Archives, I 103). Thus by mid-1718 a beginning had been made. Flamsteed reported on this enterprise in the last letter of his life, written to Sharp in November of 1719:

> Dr Halley has showed his new tables at the Temple Coffee-house: but I am told, by one that dwells in London, they are not yet finished. (Baily 1837, p. 332)

The first account of the new tables as having been completed comes in May of 1720, with Halley as the new Astronomer Royal. Crosthwaite wrote to Sharp that he was shown:

> Dr Halley's lunar tables (not yet published); but I cannot find they will give the moon's place so near the observed as Mr Flamsteed's. (Baily 1837, p. 335.)

In 1720, Pierre des Maizeaux wrote to Conti that he was expecting to receive a copy of Halley's tables:

> Nous aurons bientôt les tables astronomiques de Mr Halley, corrigées & augmentées" (*Corr.* VII p. 100, 11 Sept.).

Halley commenced his sequence of meridian-transit observations in January of 1722 (Ch. 12, II). By the spring of 1722, Delisle was advised that the Tables were almost ready for printing:

> le livre de M. Halley sur ce sujet etoit presque achevez d'imprimer. (N. Struyk to Delisle, 4 April 1722, Paris Observatory Archives, II 34.)

Delisle came to visit Halley in the summer of 1724 after which, to quote Schaffer: "Delisle's endorsement of Halley's tables was little short of ecstatic" (Thrower 1990, p. 269). Contrasting his "English" approach with that of his predecessor Phillipe de la Hire, Delisle explained:

> J'avais aussi eu soin de construire mes tables sur une théorie régulière & uniforme, tant geométrique que physique, qui était celle des Anglais.... Prêt à publier ces tables j'ai enterpris le voyage d'Angleterre, uniquement pour savoir ce qui en était des tables de M. Halley que l'on m'avait dit qu'il était sur le point de publier & qui devaient surpasser tout ce qui avait été fait jusqu'à present; & j'ai trouvé effectivement les tables de M. Halley déja imprimées depuis quelques années main non pas encore publiées. (Delisle to P. Nicasius Grammatici, Ingolstadt, Oct 1724, Paris Observatory Archives, II 128)

One copy of Halley's "unfinished astronomical tables" (Wolf 1956, p. 343) was sold in his lifetime, for a guinea, to James Logan. Logan, an Irish emigré from Pennsylvania, recalled how, when paying a visit to London in 1723: "they were lying [in] the shop of the book-dealer William Innys, at whose expense they had been set up in type, and he sold them in sheets to me privately for a very moderate price." The papers remain to this day in the Logan library in Pennsylvania (Tolles 1956, p. 25).

It thus appears that, by 1724, Halley's tables were "effectively" printed though not published. There was no hint that Halley had decided not to publish them; indeed, Delisle's non-publication of his own tables would hardly make sense in such a context. No original manuscripts of Halley's tables remain (*Corr.* VII, Note 14 on p. 101). DeLisle's *Lettres sur les Tables astronomiques de M.Halley* of 1749 reviewed this situation.

IV. LeMonnier

LeMonnier's *Institutions Astronomiques* of 1746 fully embodied the procedures of *TMM*, and is *the* publication giving Newton's sevenfold lunar theory in its most accurate form. Two modifications from the *Principia* were introduced, namely, the reversal in sign of the sixth equation, first accomplished in 1710 by Mr Cressner (Ch. 10, IX), and the adjustment of mean apse and Moon motions as in the 1726 Third Edition. LeMonnier used what we have called *TMM-2*, the optimal form of Newton's lunar theory.

Each table in LeMonnier's *Institutions* reached its maximum at the appropriate *TMM* value, with the equations in their proper sequence. Whereas Halley's tables alternated in using sometimes the 1702 constants, at other times those of 1713, LeMonnier consistently used the former. Thus his tables of the Variation reached their maximum at 37' 25", as *TMM* specified, while his text discusses the fact that in 1713 Newton gave between 37' 11" and 33' 14" as the range of this maximal value, between winter and summer. For the seventh equation, Lemonnier used 2' 10" as *TMM* had specified, while discussing the addition of 15" to that equation in 1713. The extra epicycle was discussed as an option, and the mode of constructing the additional table outlined. Their author was aware of the modified values to the equations given in 1713, contrary to what Baily stated in the above-quoted letter (pages 214–5 above).

Concerning the apse equation, LeMonnier discussed how Newton's value of 12° 18' was greater than Flamsteed's value of 11° 47', and he used the former. His mean apse position (Fig. 5.5) was one of the most accurate. LeMonnier's mean motions were more accurate than those of Halley, due to his adopting the *PNPM* Third Edition values for mean apse and Moon.

LeMonnier gave two worked examples, which differed by a mere two hours in time, and were immediately prior to a solar eclipse, of 1739. To convert *TMM-2* to LeMonnier's procedure, 1.4 and 1.7 arcminutes are added on respectively to the mean Moon and apse positions (See Ch. 5, Section V). In these two worked examples (subtracting nine minutes, twenty seconds of time from his "temps moyen" to give GMT), his seven steps of equation appear as within arcseconds of the *TMM* program, except only for the Equation of Centre, which was forty arcseconds too small; this was the case for both his worked examples.

LeMonnier's title page merely averred that "new tables" had been constructed, while its preface affirmed:

> On s'est donc appliqué uniquement à achever les Tables de la
> Lune de M.Flamsteed. (LeMonnier 1746, p. xxiv)

A later introduction to the Tables referred to the British astronomer seven times, opening as follows:

> Les Tables de la Lune que l'on donne ici sont dues principale-
> ment aux grandes découvertes que M.Newton a faites dans la

Théorie de cette Planète: on avait regardé jusqu'ici comme les meilleures celles que Flamsteed publia pour la seconde fois il y a plus de 60 ans dans le cours de Mathématique du Chevalier J. Moore; mais ces Tables "tant encore fort imparfaites, l'Auteur s'appliqua depuis à les perfectionner, en y substituant la plus grande partie de celles que l'on trouve ici. Quoique ces dernières Tables de Flamsteed n'ayent pas été publiées, on ne saurait assurer cependant si c'est uniquement parce qu'elles n'étaient pas achevées. Dans l'état où elles se trouvaient lorsqu'elles nous ont été communiquées, on jugea d'abord qu'il n'y manquait que la Table qui sert à calculer les Latitudes. (LeMonnier 1746, p. 155.)

Lemonnier here appears as knowing more about Flamsteed's post-*DOS* labours than ever did the British, in the manner of one announcing a scoop who is not at liberty to disclose his source. The Flamsteed tables, it was claimed, contained all he had required except the procedure for finding latitude.

The historian Lalande endorsed this claim:

Ces "Institutions Astronomiques" sont un des meilleurs ouvrages qu'on ait faits en Français sur l'astronomie élémentaire. On y trouve des tables de la lune de Flamsteed. (Lalande 1764, pp. 428–9.)

and Delambre wrote likewise:

il fut le confident et le continuateur de Halley et de Bradley; par ses observations, il tient a l'ecole de Piccard; par ses livres, il est de l'école de Greenwich;

adding, "les tables de la lune sont une oeuvre posthume de Flamsteed." (Delambre 1827, pp. 179, 182). This remark appears to be the sole basis for the *GHA*'s (not unreasonable) claim, that it was Halley who gave the Flamsteed tables to LeMonnier (*GHA* p. 268). D'Alembert described how, in mid-eighteenth-century France, the most widely used lunar tables were those of Halley and LeMonnier (D'Alembert 1754, I, p. iv). He noted differences between the Halley and LeMonnier tables (Ibid, III, pp. 5, 7, 10, 13, 33–35), e. g., that Halley had omitted the seventh equation while LeMonnier kept it.

Flamsteed's assistant James Hodgson inherited the Flamsteed archives through marriage to the astronomer's niece. He claimed, in the Foreword to a 1749 publication (concerning the positions of Jupiter's satellites, on which he had worked while at the Observatory), that:

> But, as to the lunar tables, the publication of them was delayed for very good reasons; and now to my very great surprise I find them printed in M. Le Monnier's *Institutions Astronomiques:* but how he came by them is to me at present a mystery…. But now, after upwards of 20 years, when it was well known that I had the original by me, and did at a convenient time intend to send them into the world according to Mr Flamsteed's own directions, it was base… [etc.] (Baily 1837, p. 704)

Yet more mysterious is the question as to why, if something resembling the set of tables published by LeMonnier were in Hodgson's hands when the *Historia Coelestis Britannica* went to print, no trace appeared therein. Hodgson made no subsequent effort to publish them.

Baily's remark that LeMonnier's tables were "evidently copied" from the unpublished Flamsteed manuscripts (Baily 1837, p. 705) is partially true, in that he had access to them. However, it is equally evident that LeMonnier wanted his tables to differ sufficiently from Flamsteed's that he could not be accused of plagiarism. His six pages of Equation of Centre are substantially those of Flamsteed, and gave columns for five different eccentricity values as did Flamsteed; however, the last two values have been altered, and so all his Equation of Centre values are different in these columns. The Flamsteed tables have pages of logarithms of the Earth-Moon distance, as are often found in tables of this period, but these are absent from LeMonnier's opus. LeMonnier's second node equation is different, to which we now turn.

V. The Node Equation

The eighteenth century saw a diminution in the amplitude of this equation under the influence of *TMM*. We saw earlier how *TMM* defined a triangle that generated the second node equation, but did not as such specify its peak value (Ch. 9, VI), and how this spawned a range of values for the function.

Table 14.2 shows how Kepler's 1627 value (Kepler 1627, p. 86) appears as generally more accurate than subsequent ones; also, that Dunthorne has somehow copied from the Flamsteed tables, while LeMonnier constructed his own node table; and that, with the exception of Cassini (1740), whose opus appeared to have no node equation table, the De la Hire textbook and Leadbetter's later 1742 opus, all the eighteenth-century textbooks here

Table 14.2: Amplitudes for the lunar node equation, i.e., the sin $2(S - N)$ term.

		TMM-Based	Others	
1627	Kepler		1° 39′ 46″	
1653	Shakerley		1° 46′ 00″	
1661	Streete		1° 45′ 00″	
1681	Flamsteed		1° 39′ 46″	(+ Whiston 1707)
1703	Flamsteed MS	1° 29′ 41″		
1716	Delisle	1° 30′ 00″		
1726	Grammatici	1° 29′ 58″		
1732	Wright	1° 29′ 45″		
1735	De la Hire	1° 34′ 00″		
1736	Leadbetter	1° 29′ 45″		
1738	Capello	1° 30′ 00″		
1739	Dunthorne	1° 29′ 41″		
1742	Leadbetter	1° 45′ 00″		
1742	LiXiang	1° 29′ 42″		
1746	LeMonnier	1° 29′ 34″		
1719/49	Halley	1° 29′ 45″		
	correct value:	1° 36′ 11″		

consulted were seeking this node equation in accord with the *TMM* instructions.

VI. Wright Claims a Longitude Prize

In 1728, *An Humble Address to the Rt Honorable Lords…relating to the Longitude* was published in London by "R.W.," claiming to contain "Sir Isaac Newton's theory freed from some errors of the Press." It contained no Tables, but had six worked examples showing *TMM*'s seven-step procedure. Both its third and sixth equations had their sign reversed. The average error of these six worked examples was 8′. He published a more extensive account in 1732, entitled *New and Correct Tables of the Lunar Motions according to Newtonian Theory* in which thirty worked examples were given, all at lunar eclipse times, a position which set equations three, five and seven to zero. The accuracy of these worked examples was ostensibly within an arcminute or two.

Baily observed that Wright was the first Briton to publish *TMM*-based tables, but then added:

> The whole however is an abortive production; for, only the second of Newton's equations is distinctly introduced; while the third & fourth seem to be wholly omitted, and the seventh united to the Variation. Moreover the maximum of eccentricity is quite at variance with Newton's assumptions. (Baily 1837, p. 701.)

Baily was evidently misled by perusing Wright's pages of worked examples for the Moon at syzygy, where several of the new equations are omitted as then having a zero value. Wright's lunar eccentricity was strangely small, reaching its maximum at 0.0619 as compared with *TMM*'s of 0.0668. Worse, he badly confused the apse equation: "And above all it is to be remembered, that in order to find the eccentricity, from the Sun's true place was subtracted, not the first aequated but the second or true place of the Moon's apogee." (Wright 1732, p. 81) Conversely, *TMM*'s text indicates that the "Annual Argument" $(A_1 - S_1)$ is represented in the *TMM* diagram by angle *STB* and not *STF* as it would be on Wright's interpretation. (Ch. 9, IV.)

VII. Leadbetter

In the view of Baily, the "more perfect adoption" of *TMM* into a tabular form was accomplished by Charles Leadbetter in his *Uranoscopia* of 1735 (Baily 1837, p. 702). The computation there presented had the usual seven steps (Leadbetter 1735, p. 84). What he then published curiously resembled Halley's then-unpublished version of *TMM*, by taking:

- the sixth equation as its fourth, though it did not incorporate Newton's modulations of the equations by seasonal etc. terms, in which respect it was a more rudimentary version.
- the 1702 constants for the annual equations and for most of the other equations, except the sixth equation, which became the fourth, for which he took the 1713 amplitude value of 2′ 25″.
- 35′ 10″ for the Variation, with no modulating term.
- the 1713 version of the maximal apse equation, viz. 12° 18′.
- the second node equation at 1° 29′ 45″.

Leadbetter surely must have obtained his procedure from Halley. On the other hand, he maintained the seventh equation, which Halley omitted, and there was a distinctive feature in his presentation: his lunar Equation of Centre used the "upper focus angle" of the ellipse, an approximate method associated with the Cambridge mathematician Seth Ward, and was set to a rather small maximal value of 4° 57' 40".

Did Leadbetter used Halley's tables without acknowledgement? Nicholas Delisle, in his *Lettres sur les tables astronomiques de M. Halley* of 1749 recalled that, in the year 1724, Halley gave him a copy of his tables after he had promised not to show them to any astronomer, and to reserve them for his private use (Baily 1837, p. 705).

In 1742 a more comprehensive two-volume textbook was published by Leadbetter, as the Second Edition of *The Compleat Astronomer*, the first edition having been in 1728. Vol. 1 claimed that its tables were "grounded upon Sir Isaac Newton's radices, and the Observations of Mr Flamsteed'; despite which, as Baily noted, all trace of the Newtonian theory has vanished, replaced by an older three-stage procedure.

In this publication, Leadbetter retained his "upper focus" equation of centre method, calling it the "evection," and what he called the "reflection" now had the maximal value of 37' 33" (As Ch. 9, V showed, the Variation had a higher value in non-Horrocksian theories. His "reflection" varies as $2(L - S)$ and is clearly the same function. Delambre's *Histoire* had a section on Leadbetter, though admitting "Cet astronome est peu connu." (Delambre 1827, p.) Concerning the tables in Leadbetter's lunar theory of 1742, given without specifying his source, Delambre averred that one was from Streete and another from Shakerley, "souvent reproduite." I could not discern these resemblances.

Each of the 1735 and 1742 Leadbetter textbooks had two worked examples, using radically different methods: the first volume using largely Halley's version of *TMM*-2 and the second his own version, of uncertain origin. Their accuracies for lunar longitude in arcminutes were as follows:

Leadbetter's worked examples compared					*long. given*	*Error*
1735	p. 88	1731	May	7	10h	6s 22° 20'1 5"	−4'.2
1735	p. 90	1735	Sep	16	noon	6s 08° 25' 32"	−2'.0
1742	p. 384	1741	Aug	28	16h2m	5s 06° 34' 59"	+26'
1742	p. 383	1740	Apr	7	noon	9s 18° 24' 04"	+16'

We are startled to see how the author of the main British astronomy textbooks during Halley's tenure as Astronomer Royal came to lose faith in the procedure of Halley, and revert to a far less accurate method. Like Brent and in the same decade, he tried and then threw off the new equations. This may remind us of how novel was the notion of adding new equations.

IX. Other Tables

The Compendious Astronomer by Charles Brent of 1741 was similar to Wright's in giving both the third and sixth equations the wrong sign, explaining: "that Author [i.e., Newton] having, by an Oversight, made the third equation additive, where it should be ablative." (Brent 1741, p. 161.) He also followed Halley and Wright in taking the sixth equation as his fourth. He not surprisingly concluded that Newton's new equations were hardly necessary, and gave several final examples omitting them. His Equation of Centre was also inaccurate, giving one or two arcminute errors. Brent's textbook was the worst of the "Newtonians." Comparing Brent's two worked examples with the *TMM-2* program showed that they averaged seven arcminutes of error (see Table 14.3, opposite).

	1729 Jan 2nd noon	*1738 Dec 12th 5.27 pm. GMT*
Brent obtained:	3s 2°20′ 15″	2s 5°24′ 16″
TMM-2-PC gives:	3s 2°25′ 59″	2s 5°27′ 23″
Correct value:	3s 2°28′ 20″	2s 5°28′ 50″

In Venice, tables were published by Angelo Capello in 1737, which followed the example of Halley and Leadbetter of omitting the seventh equation and adjusting the remaining sequence. Capello compared their accuracy with the ephemerides of Manfredius and Ghislerius, and with tables of Nicasius and De la Hire, for lunar latitude and longitude, finding his own the most accurate.

Richard Dunthorne's 1739 opus reconstructed *TMM-1*, ignoring earlier works with its claim to be the first to construct such tables. Some years later he sent a letter on the subject to the Royal Society, *A Letter Concerning the Moon's Motion*, which offered two improvements to the *TMM-2* procedure: from the data in Flamsteed's *Historia Coelestis*, Dunthorne ascertained more accurately the solar equation of centre at 1° 55′ 40″, and that the mean moon required one arcminute to be added on, which was "very nearly" that advocated in the Third Edition of the *Principia* (*Phil. Trans.* 1747, xliv pp. 412–420; Delambre 1827, p. 598). Using these amended values, Dunthorne ascertained that, for 100 eclipse times, the method predicted the lunar longitudes within 2–3 arcminutes (Ch. 11, III).

Table 14.3: Accuracy of lunar longitude computations in textbook worked examples, 1650–1750. Author and publication date plus date and local mean time for the computation are given, then "LONG." as the estimated longitude in degrees, minutes and seconds is followed by the "error" column, the divergence of these values from the "correct" modern value, with *TMM*-based values displaced to the right.

AUTHOR	Publican.	DATE	L.M.T	LONG. E	ERROR (MIN)	
Lansberg	1632	1601 11 29	12.15	081 27 45	−17	
Lansberg	1632	1602 09 26	16.59	087 24 32	−13	
Wing	1651	1587 08 17	18 33	086 22 47	−8	
Shakerley	1653	1651 05 13	23 10	119 24 52	12	
J.Newton	1657	1587 08 17	18.19	090 57 39	14	
Pagan	1658	1638 09 04	11.47	127 37 17	31	
Streete	1661	1586 09 22	14.24	067 24 24	−10	
Streete	1661	1594 12 19	15.03	133 49 36	2	
Wing	1669	1587 01 15	14 23	145 34 35	1	
Flam/Horrox	1673	1672 02 23	11.35	055 37 13	13	
Flamsteed	1681	1680 12 22	06.35	065 09 52	11	
Greenwood	1689	1594 12 19	15.03	133 48 08	−0.4	
Greenwood	1689	1586 09 22	14.24	067 26 24	−11	
P.de la Hire	1727	1704 05 15	18 37	187 40 03	−6	
Wright	1728	1692 03 18	20.55	139 41 26		+4
Wright	1728	1714 09 10	22.11	332 52 05		−10
Wright	1732	1690 06 11	12.13	273 24 15		+0
Wright	1732	1698 06 16	14.49	316 53 10		−2
Leadbetter	1735	1731 05 07	10.00	202 20 15		−4
Leadbetter	1735	1734 09 16	12.00	188 25 32		−2
M.de la hire	1735	1704 05 15	18 45	187 43 38	−3	
Capello	1737	1719 10 30	10 42	065 52 53		+0
Capello	1737	1719 11 26	16 50	065 49 05		−5
Dunthorne	1739	1737 01 02	03.40	074 08 13		−7
Brent	1741	1738 12 12	05.27	065 24 16		−5
Brent	1741	1729 02 02	12.00	092 20 15		−8
Cassini	1740	1709 11 23	12.00	145 40 01	−6	
Cassini	1740	1710 02 28	00.31	339 34 31	−2	
Leadbetter	1742	1740 04 07	12.00	288 24 04	16	
Leadbetter	1742	1741 08 28	16.02	156 34 59	−26	
Le Monnier	1746	1739 08 04	03.41	131 31 44		−1
Le Monnier	1746	1739 08 04	05.55	132 38 31		+1
Halley	1749	1684 07 02	02.41	110 52 29		−2
Halley	1749	1681 08 18	15.19	336 13 32		+1

The treatise *Nova Theoria Lunae* (Uppsala archives) by the Danish astronomer Peter Horrebov of 1718 purported to be Newtonian. Quoting from an English translation of its preface:

> The maximum equation of Apocentre Newton has established to be 12° 18′, from which the table has been computed according to the Flamsteedian method. The excentricities of the Moon have been taken over from the Horrox-Flamsteedian tables. (Source: Craig Waff, who received an English translation from Niels Jorgensen).

The six pages of tables prepared by Flamsteed in 1702/3 for the Equation of Centre involved solutions of the Kepler equation accurate to arcseconds, where both anomaly and eccentricity values varied. His achieving of such precision evidently had significance for other astronomers. Horrebov's theory was published in *Biblioteca Novissima* in Magdebourg (according to Delisle's "letter" of 1750); however, a substantial section on Horrebov in Delambre's *Histoire* (1827, pp. 140–155) made no mention of it, not did it appear in Horrebov's posthumous three-volume *Operum* (1741) so it can hardly have been widely known.

Table 14.3 compares the lunar longitude accuracies from worked examples given in textbooks 1650–1750, against a modern program, citing up to two per textbook. The dates and times given as local mean time have been cited, together with the longitude as given. Dividing these computations into the three groups of pre-1700, post 1700 non-Newtonian, and *TMM*-based, their mean errors appear as:

1) pre-1700	$1'.9 \pm 13'$	for 13 cases
2) post-1700 non-Newtonian	$-4'.5 \pm 12'$	for 6 cases
3) *TMM*-based	$-2'.7 \pm 4'$	for 15 cases

The second group remains rather small, yet this table provides evidence for a striking improvement in accuracy, even if not to the extent to which, as we have seen, the *TMM* procedure was capable of delivering. It supports Gautier's view that the Newtonian tables of this period

> surpassèrent toutes les précédents en exactitude. (Gautier 1817, p. 13.)

Several adjustments are required to convert historical LMTs into the Ephemeris Time required for this test: a calendar change between New Style and Old for most European sources; twelve hours added on, as their LMTs started from noon; for European sources, longitude-based conversion from their LMT to GMT; and a ΔT correction prior to around 1680 (Ch. 5 ,I), after which date it remains around merely ten seconds and so can be ignored. The ΔT adjustment as given in the *Explanatory Supplement* begins for the year 1620, and alters rapidly in the first half of the seventeenth century, from about two minutes for 1620 (R. Stephenson and L. Morrisson, 1984, p. 60). These ΔT values have been re-evaluated by Dr R. Stevenson (Stephenson 1995, p. 198), giving values of only about half this magnitude. He kindly provided ΔT values from 1580 (90 seconds) to 1650 (50 seconds), enabling the early worked examples of Table 3 to be computed, though uncertainties in the value of ΔT remain large over this period.

X. China

The astronomers of China required an accurate lunar theory, not for purposes of navigation as did Europe, but for the prediction of eclipses. The Chinese could not accept Copernicanism, and demanded a stationary earth, for longer than any other nation, even into the nineteenth century (Needham, Vol. 3, p. 449.); however, they were able to accept Keplerian ellipses in 1736, as describing solar and lunar motions. In the year 1610, in Peking, the Empire Astronomy Bureau failed to predict a solar eclipse, of 15 December. In contrast, astronomer Xu Guangqi (died 1633) had been able to predict it, using the Western theory of Tycho Brahe. Some astronomers therefore proposed that Western astronomical theories be translated, but this proved unacceptable. The Jesuit missionaries had to be careful about using the word Xi ("Western"), and the Emperor in his edicts would always refer to ideas as "new" in preference.

The Bureau again failed to predict the solar eclipse of 21 June 1692. These failures caused distress at the Emperor's palace, and prayers had to be said. The astronomers responsible were driven into exile. In the next century, the prediction of the solar eclipse on 15 July 1730 was successful. Old tables of 1723 were used for this, based on circular motion with epicycle and deferent, i.e., the Tycho Brahe theory.

The Catholic Jesuit missionaries Ignatius Koegler (from Germany) and Andre Pereira (Portugese), drew up an improved astronomical system called LiXiang KaoCheng HouBian (LXKCHB), in collaboration with 40 Chinese astronomers, at the order of the Emperor, from 1736 to 1742, and it was published in 1742. The Chinese name for Newton first appears in this document. Koegler later became director of Peking's Empire Astronomy Bureau for 30 years. From this body of theory, the Calendar Gui Mao Yuan Li was derived, and it endured from 1742 to 1924. This calendar was "an appendix to" LXKCHB. The year 1723 was selected as the epoch of this calendar table. This calendar is based on the 33-year pattern of leap year and is a uniquely Chinese creation. Each year the Calendar would be published as a book, and include material of agricultural significance, plus the correct days to marry, to bury the dead, and when ventures would be lucky. Lunar and solar eclipses had to be correct.

Dr Lu Dalong has discerned that LXKCHB was based on Newton's lunar "theory" of 1702, with some constants altered to accord with the *Principia* of 1713. The LXKCHB had no theory of universal gravitation, "as was in contradiction to missionary doctrine." The Jesuits did themselves believe in the heliocentric system, and may have had private conversations about the Earth moving, but it could not be stated publicly. (Not until 1859, when the fourth edition of John Herschel's *Outline of Astronomy* was translated into Chinese, did the Heliocentric system catch on. That year 1859 also saw the first translation of the *Principia*, or at least part of its Book I, into Chinese).

Competition existed between the Western Jesuits and the Chinese astronomers, as regards the propagation of Jesuit doctrine, and one who was capable of making a clear eclipse prediction could thereby gain an audience to see the Emperor. The historian Rong Zhenhua wrote, "Koegler has translated Newton's tables into Chinese," and averred that the German (Newtonian) tables by Grammaticus (1727) had been used for this purpose. No science historian prior to Dalong has described the multi-volume LXKCHB as containing the Newtonian recipe for finding the Moon's position. This is connected with the fact that the relevant texts were hardly publicly available.

We may compare the maximum-amplitude equations of LXKCHB with those of *TMM* (1702), with those published by Grammatici in 1726 (at Ingolstadt, Germany), and with the 2nd and 3rd editions of the *Principia*, as follows:

	China	*Grammatici*	*TMM*	*PNPM 1713*	*PNPM 1726*
Annual equations:					
Moon	11' 49"/50"	11' 52"	11' 49"	11' 52"	11' 51"
Node	9' 30"/29"	9' 28"	9' 30"	9' 27"	9' 24"
Varian.					
Min.(summer)	33' 14"		33' 14"	33' 40"	33' 14"
6th Eqn.	2' 25"		2' 25"	2' 10"	2' 25"

7th Eqn.	3′ 0″	3′ 0″	2′ 20″	1-2′	Omitted

Eccentricity:

Max.	66782	66782	66777	
Min.	43319	43319	43323	

These values show how the Chinese lunar method used values selected from both Grammatici and *TMM*. The amplitude of the seventh equation is clearly copied from Grammatici; while the annual equation amplitudes of Moon and node are rather derived from *TMM*.

Values of mean motions (for Sun, Moon, node & apogee) were available for midnight of the winter Solstices of 1723 and 1742, and for 45 days after the Winter Solstice of 1742. These revealed that the Chinese longitudes given in LXKCHB had 90° subtracted from their European equivalents, i.e., were measured not from zero Aries but from Zero degrees of Cancer. The mean motions are expressed as from midnight of Peking at the Winter Solstice. To convert from noon at Greenwich one adds 12 hours and subtracts (116 1/3)/15 hours, taking 116°1/3 as the longitude of Peking.

A worked example was given, specified as midnight Peking time on 45 days after the winter solstice of 1742. To convert from Newton's zero-value given in *TMM* (noon GMT on 31 December 1680, Old Style, one added 61 years to give 1741, subtracted 20 days to get the solstice in New Style (Gregorian), added 45 days, added 12 hours for midnight, and subtracted 7.75 hours for longitude conversion (116.33/15). This time-value was found to give a result agreeing within arcseconds for each step of the *TMM*-replica, showing they had achieved an exact agreement with Newton's procedure.

Chapter 15

Conclusion

Concerning the lunar observations supplied by Flamsteed to Newton in 1694, William Whewell wrote:

> And during this interval [i.e., after publication of the *Principia* in 1687], the result of the struggle depended upon the accordance of the theory with the best observations, which the Greenwich ones undoubtedly were. Upon these obserations, then, depended a greater stake in the fortune of science than was ever before at hazard. (Whewell 1836, p. 5).

A year later, he returned to the theme, this time in the context of Flamsteed's supposed reluctance to part with his observations, his comment upon what was achieved being:

> The reformation of the tables [by Newton] turned out more difficult than had been foreseen, and did not lead to any very great improvement till a later period. (Whewell 1837, p. 180)

Whewell acknowledged *TMM* as the outcome of this endeavour (p. 209), i.e., he appreciated that it worked to generate longitude positions from a given time.

This apprehension hardly reappears until a century and a half later, in the 1989 account by Curtis Wilson (*GHA* pp. 266–268). It is not to be found in *The Quest for Longitude* (Andrewes 1996), the proceedings of the 1993 Harvard longitude symposium: from the three-body problem in the *Principia*, this account moves straight to the French and German lunar theories of the latter half of the eighteenth-century (pp. 35–38), omitting the argument of the present work.

We have seen how *TMM* was a self-contained mechanism or set of procedures, devoid of theory. It should not be defined *primarily* in terms of "the reformation of tables," after all it was first used in 1710 by the Rev. Cressner to generate lunar positions, before any such tables existed. Likewise, the present work has expressed its instructions as a flow diagram written onto a computer spreadsheet.

One would be unlikely to find a history of astronomy or Newton biography which mentioned *TMM*'s sevenfold structure—even though, as Dr Waff has observed, its tabular format became widely emulated in astronomy textbooks of the first half of the eighteenth century. Its seven-stage sequence was undoubtedly its distinctive feature. Starting with mean lunar motion, one "equated" it through seven stages to reach an estimate of lunar centre in geocentric celestial longitude.

This Conclusion surveys briefly the evolution of lunar theory, as well as *TMM*'s paradoxical role in Newton's theorising about gravitation. There was never much to indicate that the practical problem of finding longitude was as such of interest to him, his primary motive remaining theoretical. Our study began with Gingerich's finding of little improvement in ephemerides accuracy through the Newtonian era (Ch.1, I), which accords with the above-quoted view of Whewell, while the last chapter concluded that *TMM*-based worked examples in textbooks were indeed substantially more accurate than their rivals. We comment finally on the arrival of "Newtonianism" into Britain in the form of Tobias Mayer's tables, which ended the career of the Horrocksian model.

Definite answers have here been given to issues that have remained conjectural over centuries, by means of computer-aided reconstruction of the past. Home computers have of late become powerful enough to contain the very accurate equations required for reconstructing the observations of past astronomers, giving a new basis for evaluating their achievement. It is a fairly recent thing that this can be done reliably: the mean motion equations that Meeus published in 1991 were significantly more accurate than those in his earlier publications, to an extent that may have been crucial for our investigation.

My ability to decode *TMM*'s instructions came about from a comparison of its diagram with Crabtree's diagram of the Horrocksian mechanism of 1642 (Figs. 7.1 and 7.2). For a couple of years I wondered as to what was the reference of space and time within which the motion as designated took place. I came to realise that these diagrams pertained to motion within a space defined by the immobility of the lunar apse. The Crabtree diagram thereby showed its eight stages unfolding over a period of thirteen months, depicting quarters and octants of solar apparent motion with regard to the lunar apse. During this period the epicycle in *TMM*'s diagram revolved *twice*. Apse and eccentricity equations were thereby generated out of phase, such that when one was at its maximum or minimum the other reached its mean position, and vice versa. Once this had been grasped, the path lay open to writing all the *TMM* "equations" as simple trigonometric functions. Finally, the exact sequence of operations, here called the steps of equation, had to be interpreted from *TMM*.

The computer model turned out to be considerably more accurate than the textbook worked examples of the period utilising *TMM*-based tables

(Ch. 14, IX). This was due in part to its lacking interpolation errors due to use of tables, primarily from the Equation of Centre: the average standard deviation of *TMM*'s errors was nearly two arcminutes in longitude, using its optimal format, whereas a collection of worked examples from textbooks using *TMM* gave errors whose standard deviation was four arcminutes.

With that qualification, Table 15.1 traces the evolution of lunar theories from Ptolemy to Mayer, comparing their accuracy (standard deviation in arcminutes) against the number of "equations" in each theory. Its first five error-estimate values were derived from computer reconstruction of the models, while sources for the last three are described below.

Astronomer	Date	No. of "Equations"	Accuracy (arcmin.)
Ptolemy	145	2	±40
Flamsteed	1681	3	±6.5
Newton	1702	7	±3.8
Halley	c.1722	6	±2.2
Lemonnier (*TMM*-2)	1745	7	±1.9
Euler	1753		±1.7
Mayer	1754	13	±0.5
Mayer	1763		±0.3

Table 15.1: Lunar Theories from Ptolemy to Mayer, AD 145 - 1763, showing number of equations versus accuracy.

Gingerich reconstructed Ptolemy's lunar theory (Gingerich 1993, pp. 60–62), showing not surprisingly that its main source of error was the then-undiscovered Variation, fluctuating by ±40 arcminutes. Two lunar equations were known in the time of Ptolemy, as remained the case in the time of Copernicus, these being the evection and equation of centre. Two further terms were discovered by Tycho Brahe and Kepler, namely the annual equation and Variation (*GHA* p. 194; Stephenson 1987, pp. 176–7), as well as the node equation. The first of these did not receive its correct physical interpretation until Flamsteed (see, e.g., Delambre 1827, p. 98). Horrocks combined the evection and equation of centre into a single mechanism, giving his theory just three stages. Flamsteed came down from the North of England with it, proclaiming it as the best lunar theory in existence.

The first edition of *PNPM* contained gravitational arguments over the known inequalities, viz. the annual equation and Variation, as well as the Horrocksian rocking apse mechanism, but it had no practical value for astronomers. Its celebrated "Moon-test" used uniform circular motion around an immobile force-centre to demonstrate the inverse-square law of gravity. This utilised the sidereal period of the Moon (27.3217 days), its monthly rotation against the stars.

In the autumn of 1694, Newton perused Flamsteed's list of errors in lunar latitude and longitude as derived from the latter's *De Sphaera* of 1681, ascertained by comparison with the astronomer's own observations of right ascension and declination. (reference???) The tabulated errors of up to eight arcminutes were almost an order of magnitude greater than those inherent in the observations, wherein Newton discerned the possibility of a new endeavour.

In the months following this visit he made at least one definite improvement, by adopting Halley's modification of the Horrocksian procedure. In January 1696, Flamsteed was informed (by Halley, presumably) that Newton had discerned six new equations (*Corr.* IV p. 192). Our inquiry did not find that these were then articulated into a synthesis: Whiteside has well characterised this period as a transition "From high hope to disenchantment." An unbridgeable gulf seemed to loom between the dynamics of an emerging gravity theory and the kinematic series of circular motions required by a lunar theory that would work. The latter was based on the *tropical* lunar month (27.3216 days) as was used in practice by astronomers.

Apparently abandoning the task, Newton departed from Cambridge to become Warden of the Mint. Later in the 1690s, expressions of pessimism were reported, and Flamsteed was prohibited from publicly announcing that he had devoted a year to preparing and rectifying his lunar observations for use by the mathematician (Ch. 1, IV). Instead, David Gregory was informed that Newton could not complete his lunar theory on account of the unwillingness of the astronomer to part with his data (Memoranda by David Gregory, July 1698, *Corr.* IV p. 277). Some years later and in Germany, Leibniz heard that "Flamsteed withheld his observations of the moon from Newton. On that account they say he has as yet been unable to complete his work on the lunar motion." (to Roemer, 1706, Westfall 1980, p. 541), a story which endures to the present day.

TMM appeared as a distinctively new synthesis, with six new equations added, four to the Moon plus one each to the node and apse, with a neo-Horrocksian formulation forming the centrepiece of its seven steps of "equation." Its sole, and rather conjectural, allusion to gravity theory lay in a couple of its equations having their amplitudes modulated by factors varying inversely as the cube of the Earth–Sun distance. In 1975 Cohen asserted of *TMM* that "the rules contained therein had been derived in a new manner from a physical theory" (Cohen 1975, p. 56), echoing Gregory's claim in his preface to *TMM* of 1702, and many others since. A year later, Whiteside argued to the contrary, pointing out that *TMM* had introduced a Ptolemaic-type equant, as had been banished from the heavens by Kepler a century earlier and resuscitated by Horrocks in the 1630s. It contained this because the mathematics of circular motion was the only means then available for describing the required inequalities.

A more recent assessment coming from Dr Craig Waff may be cited. From preliminary manuscripts for *TMM* remaining in the Portsmouth Collection at Cambridge, Dr Waff discerned

> a total absence of any evidence of theoretical deduction of these new equations. Most of the folios in this collection deal with the construction of lunar tables, the calculation of eclipses and related subjects; but there are no references to or uses of the theory of gravitation found among them. (Waff 1977, p. 70.)

In 1976, Whiteside observed that these manuscripts (ULC, Add. 3966, section 15) were assembled "mostly in wholesale confusion." This largely remains the case, and is the reason why they have not been more referred to in the present treatise. Some pertain to the 1694/5 period, while the latest are transcriptions from pages of Halley's tables, starting in June 1722. Of these which can be dated, those prior to 1700 hardly indicate use of the four new lunar equations. Surprisingly, *The Mathematical Papers of Isaac Newton* edited by Whiteside, Volumes VII and VIII, contain nothing on this subject, except for a few brief allusions in an Introduction (VII, pp. xxiv–xxviii). The *TMM*-oriented computations in these papers tended, intriguingly, to be written in English, the language in which it was first composed (Ch. 9, I), whereas those concerning the *Principia* were all in Latin.

We can almost concur with Craig Waff's view, except that, in these manuscripts, the two new annual equations to node and apse are presented as so deduced. In general, however, Dr Waff's view concurs with the conclusion here reached, that Newton did not evidently deduce his new equations from such principles. And yet, we found that these equations were all valid and not far from their optimum amplitudes.

The separate components of *TMM* entered into the *Principia* of 1713, but in a manner that required familarity with the original version to discern their coherence, at least for a modern reader. Not all were in sequence, and the context was now dynamical, no longer kinematic, with each component analysed in terms of gravity theory. This did not prevent a second epicycle from being added to the Equation of Center, a fact noted ironically by French historians . It was chiefly in this form, as given in the Second Edition of the *Principia*, that Horrocks's model spread across Europe, from Venice to Uppsala, a century after its birth, and flourished briefly.

The Second Edition's uneasy alliance between theory and practice in the lunar equations fractured in the Third Edition of 1725, with the deletion of one of the final paragraphs in the Scholium to Proposition 35 (*PNPM* p. 663). As well as a derivation of the seventh equation (amongst the last to be added prior to 1713), the paragraph contained the affirmation that lunar longitude was obtainable from these several inequalities as the aim of the exercise. In the Third Edition, the only one to appear in English, this Scholium merely discussed how certain equations of motion were derivable from an inverse-square law.

I. French Comment

The French were rather startled by the British reintroduction of epicycles, a century after Kepler had banished them from the heavens:

> "Ce qu'il y a de plus remarquable dans ce traité...adopté aujourd'hui de presque tous les Astronomes, et sur-tout par Newton, c'est que M. Machin y fait revivre les Epicycles, pour expliquer tous les mouvements et toutes les irrégularités lunaires ("De l'Orbite de la Lune dans le système Newtonian," *Histoire de l'Académie Royale des Sciences*, Paris 1746, p. 128).

John Machin composed an exposition on lunar theory for a 1729 English translation of *PNPM*.

Pierre LeMonnier treated *PNPM*'s explanation of the Moon's annual equation as his prime case-study of gravity theory, and held forth for three pages on the matter:

> L'orbite de la Lune se dilate, pour ainsi dire, plus ou moins par l'action du Soleil, selon que la Terre & la Lune se trouvent à une plus grande ou à une plus petite distance de cet Astre. (LeMonnier 1746, p. 142)

D'Alembert rather doubted whether this derivation of the the annual equation was really sound:

> Il en est quelques-unes que M.Newton dit avoir calculées par la Théorie de la gravitation, mais sans nous apprendre le chemin qu'il a pris pour y parvenir. Telles sont celles de 11' 49" qui dépend de l'équation du centre du soleil. (D'Alembert 1754, I, p. xiv.)

He was sure however that Newton had derived the Variation from gravity theory "avec beaucoup de clarté et précision." (D'Alembert 1754, Vol. 1, p. xiii.)

D'Alembert also commented on the accuracy of Newton's lunar theory. Astronomers had assumed *TMM*'s error was within two minutes of arc, he recalled, only later discovering that it could rise to five:

Ce n'a été qu'après plusieurs années qu'on s'est aperçu que l'erreur montait quelquefois à 5 minutes. (D'Alembert 1754, Préface, p. viii).

This was an underestimate as *TMM* had a twice-yearly periodic error of six or seven minutes of arc (Fig. 11.3a), plus a mean motion error of one or two arcminutes over its period of use.

The root problem was a lack of credibility in claiming to have accounted for the Horroxian rocking apse line and oscillating eccentricity by a gravity theory. Concerning the claim that the Horrocksian mechanism had been derived from gravity theory (Ch. 12, V), the French historian Bailly remarked, memorably:

Il [Newton] l'a laissé subsister comme une vraisemblance que peut faire attendre la vérité et tenir sa place." (Bailly 1779, p. 509).

He dismissed the Horrocksian theory tersely: "ce n'est point une cause physique."

II. The Silent Decades

Several decades elapsed prior to astronomers adopting the *TMM* procedure. Those in early eighteenth-century Europe who declined to do so retained a three- or four-stage method, the former based on that of Streete or Flamsteed and the latter exemplified by the French family traditions of the De la Hires and Cassinis. Notable in Britain was Leadbetter, who first advocated the Newtonian procedure with modifications due to Halley, then in 1742 renounced it in favour of a less accurate Flamsteedian procedure.

Edmond Halley was the first to use *TMM* systematically, completing his tables in the early 1720s, shortly after he became the new Astronomer Royal. Halley made one or two adjustments, simplifying *TMM* and adjusting the ordering of its equations, on the basis of having considered the matter for two decades, which as we saw rendered his procedure *less* accurate than the (corrected) 1702 version.

Around 1724, both Nicholas DeLisle in Paris and Halley at Greenwich had *TMM*-based tables ready for publication but declined to do so, for reasons not entirely evident. The view as given by, e.g., Forbes, that Halley withheld publication of his tables on account of his wish to improve them from his two decades of observations (Ch. 13, III), was rejected, as his final tables published posthumously were unaltered from those he started off with in 1722.

A copy remains of Delisle's tables in the Cassini Observatory Archives in Paris, lacking a statement of authorship. No copy remains of the "Flamsteed" tables copied by LeMonnier for his *Institutions Astronomiques* of 1749, nor have copies been located of the tables Halley used in his two decades as Astronomer Royal, even though they were eagerly copied by European astronomers. The non-publication of Nicholas DeLisle's extensive correspondence is here regrettable. Despite such gaps in the historical record, Dr Waff's thesis has been largely confirmed, that:

> Nearly all *new* lunar tables constructed during the first half of the eighteenth century utilised in some fashion his tabular theory [i.e., *TMM*] (in Cohen 1975, p. 79).

We have not improved upon his list of ten European astronomers who prepared such tables (Ch. 14, I).

III. The Saros

While the 18-year Saros cycle was known to antiquity, Halley was the first known astronomer to use it for eclipse predicton. His skill in eclipse prediction was based upon it, as DeLisle observed in his 1750 letter on the subject, which expertise became transferred to the problem of longitude. (DeLisle 1749–50) A section entitled "Saros" in Leadbetter's textbook of 1742 is presumably its first mention in an astronomy textbook where its duration was clearly specified. Leadbetter there inserted an apology to the effect that, earlier, he had

> called [it] Mr Whiston's Period; but Dr Halley assured me, that that gentleman had it from himself & desired me to let the world know so much (Leadbetter 1742, Vol. 1. p. 50),

indicating its novelty to astronomers of the period. Later, Delisle, who was in wide correspondence with other astronomers, affirmed that:

> M. Halley avoit établi cette période de 223 lunations [the Saros] ou révolutions synodiques de la Lune, que s'achevoient, suivant lui, en 18 ans, 10 ou 11 jours, 7 heures, 43 minutes, 45 secondes (letter of 1750),

adding that Halley had in 1714 "chosen" the Saros that began in 1700 to compute solar and lunar eclipses. Shortly after, D'Alembert referred to "Le période de M. Halley" as comprising 223 lunations (D'Alembert 1754, Vol.3, p. xv). While he was thus credited with discovering this period, posterity was less enthusiastic about his application of it.

Astronomers on both sides of the Channel rejected Halley's use of the Saros in an error-correction procedure. The French doubted that errors in a theory would recur in such a manner: Delambre damned the method as "useless," and worse, "ce n'etait pas selon le science" (D'Alembert 1754, p. xv; Delambre 1827, p. 282). In Britain it was widely misunderstood (Ch. 13, III). Our basis for disagreement with such experts derived from a computer reconstruction of the synchrony which Halley sought, which confirmed that his method would in fact have worked, just as he claimed.

Halley's evident failure of communication on this matter suggests more originality on his part than is normally assumed. In Halley's time it was fashionable to attribute notions to the ancients by way of conferring respectability, and his attribution of this period to antiquity seems to have been unduly successful. Neugebauer argued against the view that the Chaldeans used the Saros cycle for eclipse prediction, or that they assigned any definite meaning to the term "Saros," describing these notions as "generally accepted historical myth." He merely accepted that they used a "crude 18-year cycle" for predicting lunar eclipses, noting that "There exists no cycle for solar eclipses visible at a given place" (Neugebauer 1957, pp. 141, 142). Clearly, however, the cycle was given in Ptolemy's *Almagest* as known to the Chaldeans and as interlinking the synodic, anomalistic and draconic months (Neugebauer 1975, 1, p. 310). For a more recent discussion see North, 1994, pp. 35–47.

One of the first astronomers to be overtaken by the pace of progress, Halley's painstaking observations and his Saros technique were obsolete when published posthumously in 1749, replaced by Newtonianism arriving from the Continent.

IV. End of the Horrocksian Era

What came to be called the "Newtonian" approach rejected Newton's actual procedure. The features which came to be ascribed to him in popular accounts, such as use of second-order time differentials, and for that matter trigonometric functions of temporal variables, entered Britain in the shape of Tobias Mayer's theory of lunar motion. This was a *theory* in the modern sense, as having been derived from principles of gravitation. It had no oscillating apse line or varying eccentricity, the core of Horrocks's theory.

Mayer was posthumously awarded the longitude prize in 1760 for having accomplished what in 1683 Flamsteed viewed as unattainable. His work was influenced by the work of Leonhard Euler, whose *Theoria Motus Lunae* was published in St Petersburg in 1753. The latter's theory had maximal errors of ± 5' (Forbes, 1980, p. 142), and as such was hardly any improvement upon that of Newton half a century earlier. Remarkably, there appeared in the same year Mayer's *Novae Tabulae Solis et Lunae*, in the *Gottingen Commentarii* (Forbes 1980, p. 142). Britain's Astronomer Royal Bradley expressed the view that they did not err by more than one and a half arcminutes (letter to the Lords Commissioners of the Admiralty of April 14, 1760, Mayer 1770, p. cxi).

Britons read about Mayer's theory in the August 1754 edition of the *Gentleman's Magazine*. It was there explained how the remarkable accuracy had been obtained by use of eclipse positions separated by Saros intervals, or multiples thereof, which facilitated the checking of his mean motions. To quote Forbes, Mayer used "the method proposed by Edmond Halley for the calculation of lunar and solar eclipses" (Forbes 1980, p. 136). Where Halley used the Saros to add an irregular error-correction to *TMM*, Mayer used it for discerning the fine adjustment to his equations (Forbes, 1980, p. 136).

Mayer followed Euler in halting the oscillation of the apse and lunar eccentricity: the account described how Euler had "first substituted a constant equation of the centre, along with the evection, instead of a variable eccentricity." For a century, centred on the Newtonian endeavour, the lunar apse oscillated, since which it has retained but a single forward motion. The oscillating ellipse in the Horrocksian theory did however suggest to Euler his method for analysing the variation of orbital elements (*GHA*, p. 201).

Lastly, the author of this article (J. B., identified as John Bevis) explained that "the motion of the Moon's longitude is to be corrected by 13 equations." It had thirteen stages requiring tables, although, using "equation" in its *modern* sense, Mayer's theory contained a far larger number, a sequence of 122 such being listed in his *Theoriae Lunae* published posthumously in 1767 (Mayer 1767, pp. 23–28).

Mayer died in 1762, leaving some improvements to his theory, tables utilising which were prepared in 1763. The *Nautical Almanac* of 1767 utilised Mayer's final version of his theory, and had a standard error of ±17" in its lunar noon longitudes (Ch. 1, III). If we suppose that what historical persons meant by maximum error of a theory, was twice the standard deviation of its error, then Mayer's first theory had a standard error of around ±30", indicating the remarkable improvement he achieved between first and final formulations of his theory.

Eight decades after Newton and Gregory visited Flamsteed in the Autumn of 1694 at Greenwich, there inspecting tables of discrepancies between theoretical and observed values of lunar longitude, stimulating the mathematician to begin that endeavour which we have here examined, the great problem was resolved. The three-body problem was dealt with in a manner far beyond the scope of the present treatise. Ultimately it was a story of success, of successful endeavour: in the course of which the stars received their numbers from the first Astronomer Royal, time became measured from the setting of his clock, and longitude divisions of the globe were marked from his workplace.

Appendices

I. Some Astronomical Constants, compared with TMM

Some values such as solar eccentricity have a secular component, and the *Explanatory Supplement* of the Astronomical Ephemeris has been used to locate these, also Spencer Jones' *General Astronomy*.

		Newtonian	*Modern (for 1690)*	*Difference*
Lunar				
Eccentricity,	mean	0.055050 ±0.01173	0.5490 ±0.0117	+0.2%
Inclination,	mean	5° 08' 27"	5° 08' 43"	-16"
	max	5° 17' 20"	5° 20' 06"	
	min	4° 59' 35"	4° 57' 22"	
Apse equation		12° 18'	12° 20'	2'
Annual Equation		11° 52'	11° 16'	36'
Tropical month		27ᵈ, 7ʰ, 43ᵐ, 4ˢ.9	27ᵈ, 7ʰ, 43ᵐ, 4ˢ.7	0.2 sec.
Solar				
Eccentricity		0.016917	0.016834	+0.5%
Apse motion		1° 45'/century	1° 43'/century	
Tropical year		365d 5h 48m 57s	365d 5h 48m 45s	+10 sec
Sidereal year		365d 6h 9m 14s	365d 6h 9m 9s	+5 sec

For secular-variation term of eccentricity of Earth's orbit, see *Explanatory Supplement* p.98.

II. The Equations of Mean Motion

Epoch positions were computed for integer Julian centuries, using formulae of: Meeus 1991 (Belgium), Chapront-Touzé 1992 (France), the US-UK 1992 *Explanatory Supplemement*, and occasionally the 1961 UK *Explanatory Supplement* (of the Astronomical Ephemeris). The Chapront-Touzé computation is in arcseconds, while those of the Astronomical Ephemeris and Meeus are in degrees. Mainly, Meeus has just converted Chapront-Touzé's equations from arcseconds to degrees.

For lunar longitudes, the Meeus and Chapront-Touzé equations give comparable positions, for integer Julian century values AD 1600 to 2000. A fraction of arcsecond difference results from a speed of light correction applied by Meeus. Differences between the US-UK values and the Continental ones, amount to as much as 15 arcseconds for 1600. The latter are more up-to-date, and are more convenient, since they give the five variables as we require them, wherease the others give anomaly values.

The formulae from Meeus' *Astronomical Algorithms* (1992) supersede the formulae given in his *Astronomical Formulae for Calculators* of 1986, "which were based on the older lunar theory by Brown." (Meeus, personal communication.) For solar longitude Chapront-Touzé's table (1988, p. 346) differed from Meeus', as Chapront's table gave solar position refered to a fixed equinox of 2000, whereas tropical longitude requires the equinox of date. See also M. Chapront-Touzé and J. Chapront, 1992, p. 12 for mean equations in degrees.

The Meeus 1992 equations for mean motion are found as follows. Let *JD* be (classical) Julian Day, then the time *T* measured in Julian centuries from J2000 is given by:

$$T = \frac{JD - 2451545.0}{36525}$$

The the following Meeus expressions are all in degrees of tropical longitude, i.e., referred to the mean equinox of date.

Moon's mean longitude:
  $L' = 218.316459 + 481267.881342T - 0.0013268T^2$
Mean longitude of Sun (geometric, i.e., without aberration):
  $L = 280.46645 + 36000.76983T + 0.0003032T^2$
Longitude of mean ascending node of Moon's orbit:
  $\Omega = 125.044555 - 1934.136184T + 0.002076T^2$
Longitude of perihelion of Earth (= aphelion of "Sun"):
  $\pi = 102.93735 + 1.719526T + 0.00045962T^2$
Longitude of perigee of lunar orbit:
  $\pi' = 83.353243 + 4069.013711T - 0.0103238T^2$
Two examples of Explanatory Supplement Equations:
  $\Omega = 125° 2' 40".3 - (5^r + 134° 8' 10".5)T + 7".4T^2$
  $L' - \Omega = 93° 16' 18".9 + (1342^r + 82° 1' 3".1)T - 13".25T^2$
  $(5^r = 5 \times 360.)$

N.B.: The Julian Calendar Date corresponding to −3 Julian centuries, from the AD 2000 epoch, is December 19th, 1699 O.S. or December 29th, N.S.

III. Textbooks consulted, 1650–1750

J. B. Morinus	1650	*Tabulae Rudolfinae*
Vincent Wing	1651	*Harmonicon Coeleste*
Jeremy Shakerly	1653	*Tabulae Britannicae*
John Newton	1657	*Astronomia Britannica*
Thomas Streete	1661	*Astronomia Carolina*
J.Baptista Riccioli	1665	*Astronomia Reformata*
John Flamsteed	1681	*De Sphaera*
Phillipo de la Hire	1687	*Tabulae Astronomiae*
Nicholas Greenwood	1689	*Astronomia Anglicana*
Isaac Newton	1702	*TMM*
"	1713	*PNPM*
Nicolas Delisle	1716	*Tables du soleil & de la Lune*
Nicas. Grammaticus	1726	*Tabulae Lunares (Ingolstadii)*
Robert Wright	1732	*New & correct Tables of the lunar motions*
Angelo Capello	1738	*Astrosophiae Numericae*
Jacques Cassini	1740	*Tables Astronomiques*
Charles Leadbetter	1742	*Complete Astronomy*
Richard Dunthorne	1739	*Practical Astronomy of the Moon*
Charles Brent	1741	*The compendious Astronomer*
Pierre Le Monnier	1746	*Institutions Astronomiques*
Edmond Halley	1749	*Tabulae Astronomicae*

I couldn't find J.Hecker's *Motuum Coelestium Ephemerides* 1662, or Kirchius', *Ephemeridum Motuum Coelestum* 1681 (Lipsia), or Bealieu Desforges' *Ephemerides des mouvements celeste* 1703; Other tables had mean motions I could not fathom: Phillipe Van Lansberge's *Opera Omnia* of 1663, Maria Cunita's *Urania Propitia* of 1650, and Comte de Pagan's *Les Tables Astronomiques* of 1658.

Local Time adjustments: To obtain GMT, four minutes of time are subtracted per degree of longitude, from the local mean time.

Town Old name	Long. East	Time		Astronomers
London	−00° 05′	−00.3	min	Streete
Paris	02° 20′	09.3	"	La Hire, Cassini, Le Monnier
Zelandiae	03° 36′	14.4	"	Van Lansberge
Copenhagen	11° 07′	44.5	"	Horrebow
Bologna, Bononia	11° 20′	45.3	"	Riccioli

Ingolstadt	11° 26′	45.7	"	Grammaticus
Venice	12° 20′	49.3	"	Capello
Hven, Uraniborg	12° 45′	51.0	"	Morinus (Denmark)

IV. Mean Motions from Textbooks 1650–1750

The textbooks cited ecliptic longitudes as signs (i.e., 30° intervals), degrees, minutes and seconds, for mean positions, measured from zero Aries for noon on Dec. 31. Three sets of epoch values for the five variables are given, as used in the graphs of Chapter 5. Some tables cited these for 1680, e.g., and others, 1681. For continental values, calendar change and time adjustment have to be allowed for. In matching the modern equations to historic values, a discrepancy of a year becomes evident in apogee and node values, that of a day in solar position and that of hours in lunar longitude.

	Moon	*Sun*	*Node*	*Apogee*	*Aphelion*
Wing (1651)					
1601	0. 7. 33. 29	9. 19. 58. 34	9. 11. 35. 4	7. 19. 0. 30	3. 5. 43. 28
+40	8. 27. 7. 28	17. 57	1. 23. 40. 27	6. 7. 43. 8	0. 0. 41. 5
+60	1. 10. 41. 12	26. 56	2. 20. 30. 40	9. 11. 34. 42	1. 1. 38
+80	5. 24. 14. 56	35. 54	3. 17. 20. 54	0. 15. 26. 16	1. 22. 10
Shakerley (1653)				*anomaly*	
1620	5. 4. 15. 14	9. 21. 6. 15	-	6. 11. 16. 17	3. 5. 54. 14
1640	9. 17. 48. 57	15. 13		7. 20. 58. 19	6. 13. 11
1660	2. 1. 22. 41	24. 12		9. 0. 40. 20	6. 32. 8
Flamsteed (1681)					
1661	1. 18. 10. 14	9. 20. 25. 46	6. 21. 4. 47	5. 0. 21. 51	3. 6. 35. 0
1681	6. 1. 43. 58	34. 48	5. 24. 14. 33	8. 4. 11. 51	6. 51. 40
1701	10. 15. 17. 44	43. 50	4. 27. 24. 20	11. 8. 1. 51	7. 8. 20
Whiston (1710)					
1681	6. 1. 45. 45	9. 20. 34. 46	5. 24. 14. 35	8. 4. 28. 5	3. 7. 23. 30
1701	10. 15. 19. 50	43. 50	4. 27. 24. 20	11. 8. 18. 20	40. 10
1721	2. 28. 53. 55	52. 54	4. 0. 34. 6	2. 12. 8. 6	56. 50
Dunthorne (1739)					
1721	2. 28. 53. 55	9. 20. 52. 54	4. 0. 34. 5	2. 12. 8. 35	3. 8. 5. 30
1741	7. 12. 28. 0	21. 1. 58	3. 3. 43. 50	5. 15. 58. 50	26. 30
1761	11. 26. 2. 5	11. 2	2. 6. 53. 35	8. 19. 49. 5	47. 3

Cassini   (1740)

1720	5. 24. 29. 14	9. 10. 16. 26	4. 20. 31. 53	1. 0. 18. 26	3. 7. 56. 30
1740	10. 8. 3. 13	25. 35	3. 23. 41. 39	4. 4. 9. 17	8. 17. 5
1760	2. 21. 37. 12	34. 43	2. 26. 51. 25	7. 8. 0. 9	37. 40

Le Monnier (1746)

1720	5. 24. 30. 29	9. 10. 16. 19	4. 20. 28. 46	1. 0. 16. 51	3. 8. 4. 25
1740	10. 8. 4. 34	25. 23	3. 23. 38. 31	4. 4. 7. 5	25. 25
1760	2. 21. 38. 39	34. 27	2. 26. 48. 16	7. 7. 57. 20	46. 25

Halley (1719/49)

1701	10. 15. 20. 0	9. 20. 43. 33	4. 27. 23. 20	11. 8. 20. 0	-
1721	2. 28. 54. 5	15. 39	4. 0. 33. 5	2. 12. 10. 15	
1741	7. 12. 28. 10	28. 45	3. 3. 42. 50	5. 16. 0. 30	

V. Glossary of Terms Used in Text

Annual Argument   *TMM*'s term for the angle in zodiacal longitude between the aphelion and mean sun.

Annual Equation   Correction to be applied, using Kepler's second law or some approximation thereto, to uniform circular motion in the course of a year, e.g., of the Sun.

Anomalistic Month   27.5 days, the interval between lunar conjunctions with its mean apse.

Apogee   Furthest distance of Moon from Earth each month; this term was also used to denote aphelion, Earth's furthest distance from the Sun, in *TMM*.

Argument   Angle from which an "equation" was computed, e.g., the "annual argument" referred to the Sun-apse angle.

Apsides   The two points in the orbit of a planetary body at which it is at the greatest or least distance from the body about which it revolves. The **Apse Line** joins these two points. For an elliptical orbit (not assumed in *TMM*) it is the major axis.

Declination   Angle measured from the celestial equator and at right angles to it, taken conjointly with Right Ascension; this was replacing the older system of measuring position by celestial latitude and longitude.

Eccentricity   Conceiving an orbit as circular about a displaced centre, and taking the radius as 10^6, then eccentricity was the distance of that orbit centre from the position of the Earth about which that orbit was described; or, about the Sun instead of Earth.

Ephemeris Table showing predicted positions of a heavenly body for every day, or some multiple of days, during a given period.

Epoch for mean motion values: usually over twenty-year periods, mean motion "radix" positions were specified usually for noon of December 31st.

Equation The angle that is required to be added to a mean motion in order to "correct" it.

Equinoctial point Zero Aries in tropical longitude.

Equation of the Centre is the difference between true and mean anomalies. The principal adjustment used in *TMM*, whereby an approximation to the Kepler-equation, for a given eccentricity and apse position, was added to mean motion.

Equation of Time What Flamsteed called, "The Equation of the Naturall Days", whereby the noon "tempus apparens" deviated from uniform time, based on the Earth's uniform rotation.

"Horrox Angle" What *TMM* called the "Annual Argument" varied with the angle between the mean sun and mean apogee, here referred to as the Horrox angle. Its period is thirteen months or 411 days.

"Horrox-Wheel" The name here given to a deferent-wheel attached by Horrox to the mean apse line, whose revolution period was 6.75 months, which generated both the second apse equation and the varying eccentricity.

Julian Year, by which time in *TMM* was measured, was 365.25 days, as compared with 365.2425 days of the Gregorian calendar which Europe was then using.

Mean Anomaly Angle between planet or luminary and its mean apse.

Mean Sun This moves with uniform angular motion along the ecliptic, and coincides with the true or co-equated sun twice yearly, at the apsides.

Nodes Points where the lunar orbit intersects the plane of the ecliptic.

Obliquity of Ecliptic Inclination of the ecliptic to the celestial equator.

Octants Usually the 45° and 135° angles between the Sun and Moon, formed four times per month; or, when Newton wrote "when the moons apoge is in the Octants" it referred to the Sun-apse angle having such a magnitude.

Perigee Nearest approach of Moon each month (or of Earth to Sun, in *TMM*).

Prosthaphaeresis Term(s) to be added to an anomaly value to "co-equate" it whereby the "true" orbit position was attained.

Quadrature   90° angles between (usually) Sun and Moon, formed fortnightly.

Reduction   The transform of positions from the lunar orbit plane onto that of the ecliptic.

Right Ascension   This term is not used in *TMM*, but was the form in which data was supplied to Newton by Flamsteed, as degrees measured on the celestial equator.

Saros   A period of 223 lunations, or 11 years, 10 or 11 1/3 days (depending on leap years), when several monthly cycles closely coincided.

Sidereal   Lunar period of 27.32 days as orbit period against stars, referred to but not used in *TMM*.

Synodic   Lunar period 29.53 days, as mean period between Full Moons.

Syzygy   The conjunction and opposition of Moon and Sun, or the line joining these two positions in space.

Tropical   Year is time for Sun to return to the Vernal Equinox.

Bibliography

Adams, John Couch (1882), "On Newton's Solution of Kepler's Problem," *Monthly Notices of the Royal Astronomical Society* **43,** pp. 43–9.

Adams, John Couch (1900), *Lectures on the Lunar Theory,* Cambridge.

Andrewes, William, Ed. (1996), *The Quest for Longitude,* Harvard University Press (The Proceedings of the 1993 symposium held at Harvard).

Armitage, A. (1996), *Edmond Halley,* London

The Astronomical Almanac 1994, Washington (for Δt values).

Bailly, Jean S. (1779) *Histoire de l'Astronomie Moderne* **II,** pp. 508–10 (on Horrox), **III** (1782), p. 150.

Baily, Francis (1835a), "Some Account of the Astronomical Observations made by Dr Edmund Halley, at the Royal Observatory at Greenwich," *Memoirs of the Royal Astronomical Society* **VIII,** pp. 169–190.

Baily, Francis (1837), *An Account of the Revd. John Flamsteed, The first astronomer-Royal, compiled from his own manuscripts, and other authentic documents, never before published* (London, 1835, 1837), reprinted (Holland) 1966, with *Supplement to the Account of the Revd. John Flamsteed,* London 1837, facsimile reprinted (Holland) 1966.

Brent, Charles (1741), *The Compendious Astronomer, Containing New and Correct Tables* ("The Tables of the Moon are disposed according to Sir Isaac Newton's Theory").

Brewster, David (1855), *Memoirs of the life, Writings, and Discoveries of Sir Isaac Newton,* 2 vols., Edinburgh (reprinted Johnson Reprints 1965).

Brower, D. and Clemence, G. (1961), *Methods of Celestial Mechanics,* Ch. 12 (for node equation).

Brown, Ernest (1896), *An Introductory Treatise on the Lunar Theory,* Cambridge, 1960.

Capello, Angelo (1737), *Astrosophiae Numerica Supplementum…Exactissimae Luminarum Tabulae juxta hypotheses ac mensuras celeb. Geometrae Isaacci Newtoni,* Venice.

Cassini, Giovanni Domenico (1686), *La Connoissance des Temps,* Paris (for Equation of Time).

Cassini, Jacques (1740), *Tables Astronomiques,* Paris.

Castillejo, D. (1978) *The Expanding Force in Newton's Cosmos,* Madrid.

Chapman, Allan (1982), *Three North-Country Astronomers,* Manchester.

Chapman, Allan (1990), *Dividing the Circle, the development of critical angular measurements in astronomy 1500–1850,* New York & London.

Chapman, Allan (1990b), "Jeremiah Horrocks, The Transit of Venus, and the New Astronomy in Early Seventeenth-Century England," *Quarterly Journal of the Royal Astronomical Society* **31,** pp. 333–359.

Chapront-Touzé , Michelle, and Jean Chapront (1988), "ELP 2000–85: a semi-analytical lunar ephemeris adequate for historical times," *Astronomy and Astrophysics* **190**, pp. 340–352.

Chapront-Touzé , Michelle, & Jean Chapront (1992), *Lunar Tables and Programs from 4000BC to AD 8000,* Richmond, Virginia.

Cohen, I. Bernard (1975), *Isaac Newton's Theory of the Moon's Motion (1702), With a bibiliographical and historical introduction,* Folkestone, 1975.

Cohen, I. Bernard (1975a), "Kepler's Century," *Vistas in Astronomy* **18,** pp. 3–36.

Cohen, I. Bernard (1980), *The Newtonian Revolution,* Cambridge.

La Connoissance des Temps, see Cassini (1686).

Cook, Alan (1996), "Halley and the Saros" *Quarterly Journal of the Royal Astronomical Society* **37,** pp. 349–353.

Cook, Alan (1998), *Edmond Halley, Charting the Heavens and the Seas,* Oxford.

Colwell, Peter (1993), *Solving Kepler's Equation over three centuries,* Willman-Bell US.

Cotter, Charles H. (1968), *A History of Nautical Astronomy.*

Crabtree, William (1673): letter to Gascoigne of July 21, 1642, on Horrox theory, see Horrox *Opera,* below.

Cressener, H. (1710): "An Account of the Moon's Eclipse, February 2, 1710 observed at Streatham near London, and compared with the Calculation, by the Rev. Mr H. Cressener, FRAS" *Phil. Trans.* **27,** pp. 6–19.

d'Alembert, Jean le Ronde (1754–6), *Recherches sur différens points importans du Système du monde* Vol.1, Ch.22, & Vol.3 on history; Paris, facsimile reprint Brussels 1966.

Delambre, Jean Baptiste (1827), *Histoire de l'Astronomie au Dix-huitième Siècle,* Paris, pp. 29 (apogee & node equations), 282 (Halley's use of Saros).

Delisle, Joseph-Nicolas, *Tables du Soleil & de la Lune suivant la theorie de Mr Newton dans la 2nd Edition de ses Principles, calculées en 1716,* Observatoire de Paris Archives, MSS A2.9, No. 23.

Delisle, Joseph-Nicolas, Correspondence, Paris Observatory MSS B1.

Delisle, Joseph-Nicolas (1749–50), "Lettres sur les Tables Astronomiques de M.Halley," *Journal des Scavans,* Paris Dec. 1749, pp. 848–858, and March 1750, pp. 150–163. The second letter (which claimed Delisle was the first to prepare tables from TMM) is in the Archives of the Paris Observatoire. The first is quoted by Baily (1835, p. 705).

De Morgan, Augustus (1814) *Essays on the Life and Work of Newton*, Chicago, Open Court.

Dreyer, J.L.E. (1906), *A History of Astronomy from Thales to Kepler*, New York, 1953.

Duffett-Smith, Peter (1985), *Astronomy with your personal computer*, Cambridge (includes computer programme).

Dunthorne, Richard (1739), *Practical Astronomy of the Moon: or, new Tables... Exactly constructed from Sir Isaac Newton's Theory, as published by Dr Gregory in his Astronomy*, London & Oxford, (copy at ULC).

Dunthorne, Richard (1747), "A Letter from Mr Richard Dunthorne...concerning the Moon's Motion", *Phil. Trans.* **44** (1747), pp. 412–420, read on 5th February.

Edleston, Joseph (1850), *The Correspondence of Sir Isaac Newton and Professor Cotes*, Ed. Joseph Edleston (1850), facsimile reprint 1969.

Euler, Leonhardt (1753), *Theoria Motus Lunae exhibens omnes eius inaequalitates*, St Petersburg.

Explanatory Supplement to The Astronomical Ephemeris, HMSO, 1961; reprinted 1984, new edition 1992 by Willmann-Bell, Richmond, Virginia.

Flamsteed, John, *Lunar Tables* (manuscript), 36 pages at CUL, RGO1/50H, notebook dated 1702–1714.

Flamsteed, John (1673), *De Inaequalitate Dierum solarium Dissertatio Astronomica* London published as a supplement to Jeremiah Horrocks *Opera Posthuma* 1673, pp. 441–464.

Flamsteed, John (1674), "Extract of two Letters, written by Mr Flamsteed..." (*contra* Streete), *Phil. Trans.* **9**, pp. 219–221.

Flamsteed, John (1675), "Mr Flamsteed's letter of July 24, 1675...concerning Mr Horroxes lunar Systeme" *Phil. Trans.* **10**, pp. 368–377.

Flamsteed, John (1680) *The doctrine of the sphere, grounded on the motion of the Earth and ther antient Pythagorean and Copernican system of the World. In two parts*, London.

Flamsteed, John (1681), "Doctrine of the Sphere", in Sir Jonas Moore's *A New Systeme of the Mathematics*, 1681, Vol.1, part VI, pp. 1–75.

Flamsteed, John (1683), "A Letter from Mr Flamsteed...", *Phil. Trans.* **13** (1683) Section 154, pp. 404–8.

Flamsteed, John (1725), *Historia Coelestis Britannica*, Vols II and III.

Flamsteed, John (1975) *The Gresham Lectures of John Flamsteed* (1975), Ed. Eric Forbes.

Flamsteed, John (1982) *The Preface to John Flamsteed's Historia Coelestis Britannica,* Ed. Alan Chapman (1982), Greenwich Maritime Monograph.

Forbes, Eric (1975), *Greenwich Observatory* Volume I, "Origins and Early History 1675–1835".

Forbes, Eric (1980), *Tobias Meyer (1723–62), Pioneer of Enlightened Science in Germany,* Göttingen.

Gautier, Alfred (1817), *Essai historique sur le problème des Trois Corps,* Paris.

Gaythorpe, S.B. (1925), "On Horrocks's Treatment of the Evection and the Equation of the Centre," *Monthly Notices of the Royal Astronomical Society* **LXXXV**, pp. 858–865.

Gaythorpe, S.B. (1957), "Jeremiah Horrocks and his 'New Theory of the Moon'," *Journal of the British Astronomical Association* **67**, pp. 134–44.

Gingerich, Owen and Welther, Barbara (1974), "Note on Flamsteed's Lunar Tables," *British Journal for the History of Science* **7**, pp. 257–8.

Gingerich, Owen and Welther, Barbara (1983), *Planetary, Lunar and Solar Positions AD 1650–1805,* Harvard; see Preface, "The Accuracy of Historical Ephimerides", pp. i–xxiii.

Gingerich, Owen (1993), *The Eye of Heaven: Ptolemy, Copernicus and Kepler,* New York.

Godfray, Hugh (1871), *An Elementary Treatise on The Lunar Theory.*

Grammaticus, Nicasius (1726), *Tabulae Lunares ex Theoria et Mensuris Geometrae Celeberrimi domini Isaaci Newtoni,* Ingolstadt.

Greenberg, John (1984), "Degrees of Longitude and the Earth's Shape: The Diffusion of a Scientific Idea in Paris in the 1730s" (section 3, "Delisle's lost Chance"), *Annals of Science* **41**, pp. 151–158.

Greenwood, Nicholas (1689), *Astronomia Anglicana.*

Gregory, David (1702), *Astronomiae Physicae at Geometricae Elementa,* with 'Lunae Theoria Newtoniana' on pp. 332–336; reprinted 1726; translated as *The Elements of Physical and Geometrical Astronomy* 1715, 2nd Edition 1726, 2 vols, with *TMM* on pp. 563–571; facsimile reprint (Sources of Science, no. 119) NY & London, Johnson Reprint 1972.

Halley, Edmond, *Astronomical Notebook,* fair manuscript copy 518 pp., in Royal Astronomical Society Library.

Halley, Edmond (1679) *Catalogus Stellarum Australium* ('Quaedam lunaris theoria emendationem spedantia' p. 12).

Halley, Edmond (1697), "The True Theory of the Tides…," *Phil. Trans.* **19**, pp. 445–57.

Halley, Edmond (1708), *Foreword to* Miscellanea Curiosa (on accuracy of *TMM*).

Halley Edmond, Ed. (1712), *Historia Coelestis Libri Duo*, 'by John Flamsteed'.

Halley, Edmond (1731/2), "A Proposal of a Method for Finding Longitude at Sea within a Degree, or Twenty Leagues", *Phil. Trans.* **37**, pp. 185–95.

Halley, Edmond (1749), *Tabulae Astronomicae, accedunt de usu tabularum preacepta*, London; trans: *Astronomical Tables with Precepts both in English & Latin for comparing the places of the Sun, Moon etc.*, 1752.

Heckart, B. (1984), *Edmond Halley*, London

Horrox, Jeremiah, *Philosophical Exercises*, The Second'Part, ULC, RGO 1.68B.

Horrox, Jeremiah (1673), *Opera Posthuma*, London. Epilogue by Flamsteed p. 491, edited by John Wallis (esp. J. Horrox letter to W. Crabtree December 20th 1638, for "germ of new theory:" Wallis erroneously printed the year as 1628).

Hodgson, William (1750), *Theory of Jupiter's Satellites*, Introduction (on LeMonnier publishing Flamsteed's tables).

Horrebov, Peter (1718), *Nova Theoria Lunae* was published in "Biblioteca Novissima,' Magdeburg, according to Delisle.

Howse, Derek (1980), *Greenwich Time and the Discovery of Longitude*, London

Howse, Derek (1990), "Newton, Halley and the Royal Observatory", in Thrower (1990)

Howse, Derek (1993), "The Astronomers Royal and the Problem of Longitude" *Antiquarian Horology*, Autumn, pp. 43–52.

Howse, Derek (1994), "Longitude-finding by Astronomy" *Companion Encyclopaedia of the History and Philosophy of the Mathematical Sciences*, I. Grattan-Guinness Ed., **II**, pp. 1134–8.

Howse, Derek (1997), "The Lunar-distance Method of Measuring Longitude", in Andrewes (1996).

Hoyle, Fred (1997), *On Stonehenge*, London

Hughes, D.W., Yallop, B.D., & Hohenkerk, C.Y. (1989), "The Equation of Time," *Monthly Notices of the Royal Astronomical Society* **238**, pp. 1529–1535.

Improved Lunar Ephemeris 1952–1959 by the Nautical Almanac Offices of the USA and UK, Washington (1954); for improvement to ILE see: "Explanation, Moon", *The Astronomical Ephemeris*, Washington & London 1972, p. 539.

Jorgenson, Niels (1974), "On the Moon's Elliptic Inequality, Evection and Variation and Horrox's 'New Theory of the Moon'" *Centaurus*, **18** pp. 316–18.

Kelly, John (1991), *Practical Astronomy during the Seventeenth Century: Almanack-makers in America and England*, Harvard Dissertations in the History of Science, New York.

Kepler, Johannes (1627) *Tabulae Rudolphinae*, Prague, in *Johannes Kepler Gesammelte Werke* **10**, Munich 1969.

Kollerstrom, N. (1985), "Newton's Lunar Mass Error" *Journal of the British Astronomical Association* **95**, pp. 151–3.

Kollerstrom, N. (1990), "The Edmond Halley 'Bull's Eye' Enigma" *Journal of the British Astronomical Association* **100,** p. 7.

Kollerstrom, N. (1991), "Newton's two 'Moon-tests'," *British Journal for History of Science* **24**, pp. 369–72.

Kollerstrom, N. (1995), "A Reintroduction of Epicycles: Newton's 1702 Lunar Theory and Halley's Saros Correction" *Quarterly Journal of the Royal Astronomical Society* **36**, pp. 357–368.

Kollerstrom, N. and B. Yallop (1995), "Flamsteed's Lunar Data 1692-5, sent to Newton" *Journal for the History of Astronomy* **XXVI**, pp. 237–246.

Lansberge, Philip van (1632), *Tabulae motuum coelestium perpetuae*, Middelburg.

Lansberge, Philip van (1663), *Theoriae Motuum Coelestum*, in *Opera Omnia*, Zelandiae.

Leadbetter, Charles (1735), *Uranoscopia, or the Contemplation of the Heavens… Also, an Explanation and Demonstration of the Keplarian and Flamsteedian Methods of Computing the Times and Principal appearances of Solar Eclipses.*

Leadbetter, Charles (1742), *A Complete System of Astronomy*, 2nd edition, 2 vols.

"Mayer's new Tables of the Sun and Moon" (1754), *Gentleman's Magazine*, (August), London, **XXIV** pp. 374–376, no name; extract quoted in Forbes 1980, pp. 143–146, citing author as John Bevis.

Mayer, Tobias (1753), *Novae Tabulae Solis et Lunae, Göttingen Commentarii.*

Mayer, Tobias (1767), *Theoria Lunae juxta Systema Newtonianum,* London.

Mayer, Tobias (1770), *Tabulae Motuum Solis et Lunae,* London (contains letters of Bradley, and review of Halley's method).

McNally, Derek (1974), *Positional Astronomy,* London.

Mercator, Nicholas (1676), *Institutionum Astronomicarum, Libri Duo cum Tabulis Tychonianis,* London

Meeus, Jean (1983), *Astronomical Tables of the Sun, Moon and Planets,* Richmond, Virginia.

Meeus, Jean (1985), *Astronomical Formulae for Calculators* Richmond, Virginia, US, 3rd Edn.

Meeus, Jean (1991), *Astronomical Algorithms*, Richmond, Virginia, US.

Le Monnier, Pierre (1746), *Institutions Astronomiques, ou leçons elementaires d'astronomie,* Paris.

More, Louis T. (1934), *Isaac Newton: A Biography*, NY & London, facsimile reprinted New York, 1962.

Needham, Joseph (1959), *Science and Civilisation in China, III: Mathematics and the Sciences of the Heavens and the Earth*, Cambridge.

Neugebauer, Otto (1957), *The Exact Sciences in Antiquity*, Brown University Press, printed in Denmark (Halley's use of term 'Saros', p. 142).

Neugebauer, Otto (1975), *History of Ancient Mathematical Astronomy* Vol. I, Springer-Verlag Berlin & New York.

Newton, Isaac, manuscripts on *Comparison of the calculated place of the Moon with Observation*, (c.1694–1724, but not in chronological sequence), ULC Add. 3966,15.

Newton, Isaac, *PNPM, Isaac Newton's Philosophiae Naturalis Principia Mathematica*, 2 Vols., ed. A. Koyré and I. B. Cohen, Cambridge Massachusetts 1972.

The Correspondence of Sir Isaac Newton and Professor Cotes, Ed. Joseph Edleston (1850), facsimile reprint 1969.

The Correspondence of Isaac Newton, Volume IV 1694–1709, J. F. Scott Ed. (1967), Cambridge.

TMM: John Harris's *Lexicon Technicum* under heading 'Moon' (English) 1704, 1708, 1716, 1725, 1736; has *TMM* with quotes from Flamsteed.

TMM: *A New and most Accurate Theory of the Moon's Motion; Whereby all her Irregularities may be solved, and her Place truly calculated to Two Minutes. Writtenby That Incomparable Mathematican Mr. Isaac Newton* (English pamphlet, no name) 1702;

TMM: Miscellanea Curiosa **1** pp. 270–281 (English) 1705, 1708, 1726;

TMM: William Whiston, *Praelectiones Astronomicae* (Latin) 1707; *Astronomical Lectures* 1715, 1728, 1782 (English);

TMM: Samuel Horsley Ed., *Opera* of Newton (Latin) 1782.

Review of *TMM* in: *Histoire des Ouvrages des Savans*, January–March 1703, pp. 121–23.

"De l'Orbite de la Lune dans le syst me Newtonian", *Histoire de l'AcademieRoyale des Sciences*, **173**, Paris (1743, Pub. 1746) pp123–129.

Journal Book of the Royal Society, XII, 1720–26 (pp. 10–12, 12 May 1720 report of exchange between Newton and Halley).

Newton, Isaac (1700), *Theory of the Moon* Royal Society MS 247, ff. 15/16, copy by David Gregory, 4 sides written on two pages, dated 27 Feb then 25 Mar; original at ULC, Add.3966,10.

Newton, Isaac (1713), *Philosophiae Naturalis Principia Mathematica*, 2nd Edn.

Newton's Principia (1729), Motte translation of 3rd Edition (1725), F. Cajori edn., Vol II, University of California Press, 1962.

Newton, John (1657), *Astronomia Britannicae.*

North, John (1994), *The Fontana history of Astronomy and Cosmology*, London.

Riccioli, G.B. (1665), *Astronomia Reformata tomi duo*, Bononia (Bologna).

Roy, A.E. (1978), *Orbital Motion*, London

Royal Greenwich Observatory, *NAO Technical Note No 48, Approximate Lunar Co-ordinates* (1979), B. Emerson (gives modern equations).

Sadler, D.H. (1976), "Lunar Distances and the Nautical Almanac" *Vistas in Astronomy* **20**, pp. 113–121.

Shakerly, Jeremy (1653), *Tabulae Britannicae*, London

Smart, William (1977), *Textbook on Spherical Astronomy* Cambridge, 6th Edition.

Sobel, Dava (1995), *Longitude: The True Story of a Lone Genius Who Solved the Greatest Scientific Problem of His Time*, London.

Spencer Jones, Harold (1961), *General Astronomy* 4th Edn., London, esp. pp. 123–130 on lunar theory.

Stephenson, Bruce (1987), *Kepler's Physical Astronomy*, Springer-Verlag, Berlin & New York.

Stephenson, Richard (1834), *Newton's Lunar Theory Exhibited Analytically*, Cambridge.

Stephenson, F. R., and Morrisson, L. V. (1984), "Long-Term Changes in the Rotation of the Earth: 700BC to AD 1980" *Phil. Trans.* Ser. A, **313**, pp. 47–70.

Stephenson, F. R. (1995) "Long-Term fluctuations in the Earth's Rotation: 700 BC–AD 1990" *Phil. Trans.* Ser. A, **351** pp. 165–202.

Streete, Thomas (1661), *Astronomia Carolina, a new theorie of Coelestial Motions; An Appendix to Astronomia Carolina*, 1674; Latin translation by Morino, with added *Tabulas Rudolphinas Johann Baptista Morino*, Nürnberg 1705; 3rd Ed. (1716) (posthumous), ed. Halley, with Halley's 1680s sextant observations and "a proposal how to find the Longitude"; 3rd edition.

Streete, Thomas (1674), *The Description and Use of the Planetary Systeme together with Easie Tables*, London. A reply to Flamsteed's allegation of plagiarism over the Horroxian theory is appended.

Streete, Thomas (1716), see Streete (1661).

Taylor, E. (1956) *The Haven-Finding Art*, London.

Thoren, Victor (1974), "Kepler's Second Law in England" *British Journal for History of Science* **7**, pp. 243–256.

Thrower, Norman J. (1981) *The Three Voyages of Edmond Halley in the "Paramore", 1698–1701*, Norman J.Thrower Ed., London.

Thrower, Norman J., Ed. (1990), *Standing on the Shoulders of Giants, A Longer View of Newton and Halley*, Berkeley & Oxford; esp. Schaffer, S., "Halley, DeLisle and the Making of the Comet," pp. 254–298.

Tolles, Frederick (1956), "Philadelphia's First Scientist, James Logan" *Isis* **XLVII**, pp. 20–30.

Waff, Craig (1977), "Newton and the Motion of the Moon, an Essay Review"*Centaurus* **21**, pp. 64–75.

Westfall, Richard (1973), "Newton and the Fudge Factor," *Science* **CLXXIX**, pp. 751–58.

Westfall, Richard (1980), *Never at Rest, A Biography of Isaac Newton*, Cambridge.

Whewell, William (1836), "Newton and Flamsteed. Remarks on an article in No. CIX Quarterly Review", *Philosophical Magasine* VIII pp. 139–147 (anonymous book review), reprinted in Royal Society "Tract 46" pp. 3–19.

Whewell, William (1837), *History of the Inductive Sciences, from the earliest to the present times*, London, **I** p. 457 (Horrox), **II**, pp. 178–186, 214–222, 302–304.

Whiston, William (1707), *Praelectiones astronomicae Cantabrigae in scholiis publicis habitae…*, containing *Tabulae plurimae astronomicae Flamstedianae correctae*; plus commentary, pp. 309–327, with date Dec. 6, 1703; trans. *Astronomical Lectures, Read in the Publick Schools at Cambridge* 1715, 2nd Edition Corrected 1728, with *TMM* on pp. 345–368; Facsimile Reprint (Sources of Science, no. 122) N.Y & London 1972.

Whiston, William (1728), see Whiston (1707)

Whiteside, Derek T. (1684), *The Mathematical Papers of Isaac Newton*, **IV** (1674–1684), p. 665 for treatment of upper-focus and mean anomaly; **VI** pp. 509, 519 forHorroxian theory; **VII** (1691–1695), pp. xxiv, xxv, and xxviii, CUP 1967.

Whiteside, Derek T. (1976) "Newton's Lunar Theory, From High Hope to Disenchantment" *Vistas in Astronomy* **19**, pp. 317–28.

Willmoth, Francis, (Ed.) (1997) *Flamsteed's Stars, New Persectives on the life and work of the first Astronomer Royal (1646–1719)*, Woodbridge, Suffolk.

Wilson, Curtis, (1978) "Horrox, Harmonies and the Exactitude of Kepler's Third Law" *Science and History: Studies in Honour of Edward Rosen, Studia Copernicana*, **16,** Wroclaw, pp. 235–59, also in Wilson 1989.

Wilson, Curtis (1980), "The Perturbation of Solar Tables from Lacaille to Delambre: the Rapprochement of Observation and Theory" *Archive for History of Exact Sciences* **22**, Part I pp. 53–188, Part II pp. 189–304.

Wilson, Curtis (1985), "The Great Inequality of Jupiter and Saturn: from Kepler to Laplace" *Archive for History of the Exact Sciences* **33**, pp. 15–290 (p. 16, Flamsteed's accuracy).

Wilson, Curtis (1987), "On the Origin of Horrocks's Lunar Theory" *Journal for the History of Astronomy* **xviii**, pp. 77–94.

Wilson, Curtis, & Taton, Rene, Ed. (1989), *The General History of Astronomy, Volume 2 Planetary astronomy from the Renaissance to the rise of astrophysics,* Cambridge. Part A: Tycho Brahe to Newton, Chapters 10 and 12: "Predictive Astronomy in the century after Kepler" and "The Newtonian achievement in astronomy."

Wilson, Curtis (1989), *Astronomy From Kepler to Newton*, Variorum Reprints , London. (NB Ch. 7, "On the Origin of Horrocks Lunar Theory").

Wilson, Curtis, "Newton on the Moon's Variation and Apsidal Motion" (MS in preparation).

Wing, Vincent (1649), *Urania Practica, with divers rules and tables of extraordinary use in navigation*, London.

Wing, Vincent (1651), *Harmonicon Coeleste, conteining an absolute and entire Piece of Astronomie*, London.

Wing, Vincent (1669), *Astronomia Britannica*, London.

Wolf, Edwin (1956), "The Romance of James Logan's Books" *William and Mary Quarterly* **XIII**, pp. 342–353.

Woodhouse, R. (1812) *An Elementary Treatise on Astronomy*, Vol.II, London.

Wright, Robert (1732), *New & Correct Tables of the lunar motions, according to the Newtonian Theory*, London.

Index

Accuracy of TMM 45, 109

Annual Argument 78

Annual equations 114, 130, 230

Anomaly, mean 71

Anomalistic month 128, 153, 157

Aphelion 52, 71, 142

Apogee 28, 34, 53

Apparent time 25

Apse equation 34, 119

Apse line 27, 83

Argumentum annuum, *see* Annual
  Argument

Baily, Francis 17, 62, 139, 193, 207

Bailly, M. 29, 231

Baricentre 179

Board of longitude *see* Longitude,
  Board of

Brahe, Tycho 72, 227

Brent, Charles 218

Capello, Angelo 63-4, 218, 239-40

Cassini, Jean (solar equation) 73

Chapront-Touzé, Michel, and Jean
  Chapront 58, 237–8

Clairaut, Alexis 28

Cloudsley Shovell, *see* Shovell,
  Cloudsley

Cohen, I. Bernard 17, 175

Computer program 98

Connoissance des Temps 22, 24, 58

Cotes, Roger 166, 172, 180

Crabtree, William 81, 89

D'Alembert, J. R. 196, 202, 230

Delisle, Nicolas 208, 232

Doctrine ofthe Sphere (Flamsteed)
  76, 145

Dunthorne, Richard 94, 99, 102,
  218

Eccentricity 36, 83

Eclipse, prediction of 41, 150

Elliptic orbit 32

Ephemeris time 59

Epicycle, in *TMM* 170, 179, 230

Epoch, of *TMM* 43

Equation 72

Equation of Centre 71, 74, 77, 120

Equation of Time 23, 89

Error-envelope 48, 154, 181

Euler, Leonhardt 38, 234

Evection 85

Flamsteed, John
  Tables of, 66, 106, 151, 205, 213
  Annual Equation 131

Forbes, Eric 93, 195

Gascoigne, William 142

Gaythorpe, S. B. 91

Gingerich, Owen 19, 22

Grammatici, Nicolaus 208

Gravity theory 27, 116, 130

Greenwich meridian 25

GMT 23

Godfray, Hugh 94

Gregory, David 21, 43

Halley, Edmond 21, 24, 45, 90, 110,
  168

Halley's Ode 31
Hevelius, Joannes 51
Historia Coelestis 42
Hodgson, James 76, 214
Horrebov, Peter 220
Horrocks, Jeremiah 26, 28
Horrocks-Wheel 85, 118, 199
Horroxian theory 69, 81
Horroxian year 34, 78
Howse, Derek 25
Julian Date 59, 136
Julian years 62
Jupiter, moons of 24, 108
Kepler equation 71, 73
Kepler ellipse 38
Kepler motion 32
Kepler's Second Law 76
Latitude, lunar 128, 159
Lacaille, N. 24
Laplace, Pierre Simon 19
Leadbetter, Charles 30, 37, 216
Leibniz, G. W. 228
LeMonnier, Pierre 66, 126, 207, 211
Longitude, Board of 20, 24
Longitude, finding of 187
Logan, James 211
Lunar mass estimate 179
Mayer, Tobias 233
Master of the Mint 43, 143
Mean anomaly 72
Mean motions 56, 57, 61, 99, 237
Mean time 23
Meeus, Jean xviii, 58, 237–8
Nautical almanac 25, 234

Nautical Almanac Office 49
Newtonian dynamics 27, 165
Node equation 101, 117, 125, 177, 214
Octants 83, 87
Paraiba, longitude of 24
Perigee 34
Perihelion 71
Principia,
   First Edition 31
   Lunar section 91, 165–182, 229
   Second Edition 165–182
   Third Edition 178
Prosthaphaeresis 78
Ptolemy, Claudius 227
Reduction, to ecliptic 124
Rudolphine Tables 65
Saros xx, 26, 183, 189, 210, 232
Semiannual equation 34
Second equation 117
Seven steps 105, 225
Seventh equation 176
Sharp, A. 206, 208
Shovell, Cloudsley 20
Sidereal framework 55
Sidereal month 23, 228
Sixth equation (of *TMM*) 123, 137, 174
Sobel, Dava xvii
Solar equation 77
Steps of Equation (of *TMM*) 100
Streete, Thomas 54, 69, 146
Synodic month 157
Syzygy 197
Theory xvii

Tropical framework 168, 228

Tropical month 55, 69

Universal time 59

Uraniborg 51

Variation 32–3, 121, 166

Waff, Craig 17, 97, 168, 182, 229

Westfall, Richard S. 128

Whewell, William 57, 168, 225

Whiston, William 18, 36, 38, 61, 83, 123

Whiteside, D.T. xix, 29, 46, 84, 92

Wilson, Curtis A. xi–xiii, 134

Wing, Vincent 23

Wright, R. 215